Rob

Quantum Field Theory

1971 Lecture Notes

MINKOWSKI
Institute Press

Robert Geroch
Enrico Fermi Institute
University of Chicago

Cover: Lecture notes are often written in similar environments

ISBN: 978-0-9879871-9-8 (softcover)
ISBN: 978-0-9879871-0-5 (ebook)

Minkowski Institute Press
Montreal, Quebec, Canada
http://minkowskiinstitute.org/mip/

For information on all Minkowski Institute Press publications visit our website
at http://minkowskiinstitute.org/mip/books/

Preface

This publication of Robert Geroch's course notes on quantum field theory is the second book in the new *Lecture Notes Series* of the *Minkowski Institute Press*. The idea of this series is to extend the life in space and time of valuable course notes in order that they continue to serve their noble purpose by bringing enlightenment to the present and future generations.

Although written in 1971 Geroch's lecture notes are still a very helpful text on quantum field theory since they contain a concise exposition of its core topics accompanied by compressed but deep and clear explanations. What also makes this book a valuable contribution to the existing textbooks on quantum field theory is Geroch's unique approach to teaching theoretical and mathematical physics – the physical concepts and the mathematics, which describes them, are masterfully intertwined in such a way that both reinforce each other to facilitate the understanding of the most abstract and subtle issues.

Robert Geroch would like to thank Michael Seifert for producing the initial LaTeXversion of the typed course notes from mimeographed originals.

Montreal, February 2013   *Vesselin Petkov*
                                                     Minkowski Institute Press

Contents

1. The Klein-Gordon Equation

We want to write down some kind of a quantum theory for a free relativistic particle. We are familiar with the old Schrödinger prescription, which more or less instructs us as to how to write down a quantum theory for a simple, nonrelativistic classical system. The idea is to mimic as much at that prescription as we can. In doing this, a number of difficulties will be encountered which, however, we shall be able to resolve. There is a reasonable and consistent quantum theory for a free relativistic (spin zero) particle.

Recall the Schrödinger prescription. We have a classical system (e.g., a pendulum, or a ball rolling on a table). The manifold of possible instantaneous configurations of this system is called *configuration space*, and points of this manifold are labeled by letters such as x. However, in order to specify completely the state of the system (i.e., in order to give enough information to uniquely determine its future evolution), we must specify at some initial time both its configuration x and its momentum p. The collection of such pairs (x, p) is called *phase space*. (More precisely, phase space is the cotangent bundle of configuration space.) Finally, the dynamics of the system is described by a certain real-valued function on phase space, $H(x, p)$, the *Hamiltonian*. The time-evolution of the system (i.e., its point in phase space) is given by Hamilton's equations:

$$\frac{\mathrm{d}}{\mathrm{d}t}x = \frac{\partial}{\partial p}H \qquad \frac{\mathrm{d}}{\mathrm{d}t}p = -\frac{\partial}{\partial x}H \tag{1}$$

Thus, the complete dynamical history of the classical system is represented by curves (solutions of Eqn. (1)), $(x, p)(t)$, in phase space. (More precisely, by integral curves of the Hamiltonian vector field in phase space.)

The state of the corresponding quantum system is characterized not by a point in phase space as in the classical case, but rather by a complex-valued function $\psi(x)$ on configuration space. The time-evolution of the state of the system is then given, not by Eqn. (1) as in the classical case, but rather by the Schrödinger equation

$$i\hbar\frac{\partial}{\partial t}\psi = H\left(x, -i\hbar\frac{\partial}{\partial x}\right)\psi \tag{2}$$

where the operator on the right means "at each appearance of p in H, substitute $-i\hbar\frac{\partial}{\partial x}$". (Clearly, this "prescription" may become ambiguous for a sufficiently complicated classical system.) Thus, the complete dynamical history of the

1

system is represented by a certain complex-valued function $\psi(x,t)$ of location in configuration space and time.

We now attempt to apply this prescription to a free relativistic article of mass $m \geq 0$. The (4-)momentum of such a classical particle, p^a, satisfies $p^a p_a = m^2$. (Latin indices represent (4-)vectors or tensors in Minkowski space. We use signature $(+, -, -, -)$.) Choose a particular unit (future-directed) timelike vector t^a (a "rest frame"), and consider the component of p^a parallel to t^a, $E = p^a t_a$, and its components perpendicular to t^a, $\vec{p}$. Then, from $p^a p_a = m^2$, we obtain the standard relation between this "energy" and "3-momentum":

$$E = \left(\vec{p} \cdot \vec{p} + m^2\right)^{1/2} . \tag{3}$$

(Here and hereafter, we set the speed of light, c, equal to one.) The plus sign on the right in Eqn. (3) results from the fact that p^a is a future-directed timelike vector. It seems natural to consider Eqn. (3) as representing the "Hamiltonian" for a free relativistic particle. We are thus led to consider the dynamical history of the quantum particle as being characterized by a complex-valued function $\phi(x^a)$ on Minkowski space (x^a represents position in Minkowski space — it replaces both the "x" and "t" in the Schrödinger theory), satisfying the equation:

$$i\hbar \frac{\partial}{\partial t}\phi = \left[-\hbar^2 \nabla^2 + m^2\right]^{1/2}\phi \tag{4}$$

The first set of difficulties now appear. In the first place, it is not obvious that Eqn. (4) is in any sense Lorentz invariant - that it is independent of our original choice of p^a. Furthermore, it is not clear what meaning is to be given to the operator on the right side of Eqn. (4): what does the "square root" of a differential operator mean? Both of these difficulties can be made to disappear, after a fashion, by multiplying both sides of Eqn. (4) by another, equally obscure, operator, $i\hbar\frac{\partial}{\partial t} + \left[-\hbar^2 \nabla^2 + m^2\right]^{1/2}$, and expanding using associativity. The result is the *Klein-Gordon* equation:

$$\left(\Box + \frac{m^2}{\hbar^2}\right)\phi = 0, \qquad \text{or} \qquad \left(\Box + \mu^2\right)\phi = 0 \tag{5}$$

which is both meaningful and relativistically invariant. (We set $\mu = m/\hbar$.) We might expect intuitively that the consequence of multiplying Eqn. (4) by something to get Eqn. (5) will be that the number of solutions of Eqn. (5) will be rather larger than the number of solutions of Eqn. (4) (whatever that means.) As we shall see later, this intuitive feeling is indeed borne out.

To summarize, we have decided to describe our quantized free relativistic particle by a complex-valued function ϕ on Minkowski space, which satisfies Eqn. (5).

Just for the fun of it, let's look for a solution of Eqn. (5). We try

$$\phi = e^{ik_a x^a} \tag{6}$$

where k_a is a constant vector field in Minkowski space. Substituting Eqn. (6) into Eqn. (5), we discover that (6) is indeed a solution provided

$$k^a k_a = \mu^2 \tag{7}$$

i.e., provided k_a is timelike with norm μ.

In the Schrödinger prescription, the wave function ψ has a definite and simple physical interpretation: $\psi\psi^*$ represents the probability contribution for finding the particle. What is the analogous situation with regard to the solutions of the Klein-Gordon equation? We know, e.g., from electrodynamics, that what is a "density" in a nonrelativistic theory normally becomes "the time-component of a 4-vector" in a relativistic theory. Thus, to replace "$\psi\psi^*$", we are led to look for some 4-vector constructed out of a solution of the Klein-Gordon equation. This suggestion is further strengthened by the observation that for a Schrödinger particle in a potential (so $H = (1/2m)p^2 + V(x)$), we have the equation

$$\frac{\partial}{\partial t}(\psi\psi^*) = -\vec{\nabla} \cdot \left[\frac{\hbar}{2mi} \left(\psi^*\vec{\nabla}\psi - \psi\vec{\nabla}\psi^* \right) \right] \tag{8}$$

(Proof: evaluate the time-derivatives on the left using (2), and verify that the result is the same as the expression on the right.) This looks very much like the nonrelativistic form of the statement that the 4-divergence of some 4-vector vanishes. Hence, we want to construct some divergence-free 4-vector from solutions of the Klein-Gordon equation. One soon discovers such an object which, in fact, looks suggestively like the object appearing in Eqn. (8):

$$J_a = \frac{1}{2i} \left(\phi^*\nabla_a\phi - \phi\nabla_a\phi^* \right) \tag{9}$$

Note that, because of (5), J_a is divergence-free.

We cannot interpret the "time-component" of Eqn. (9) as a probability density for the particle unless this quantity is always nonnegative, that is to say, that $J_a t^a \geq 0$ for every future-directed timelike vector t^a, that is to say, unless J_a itself is future-directed and timelike. To see whether this is indeed the case, we evaluate J_a for the plane-wave solution (Eqn. (6)), and find:

$$J_a = k_a \tag{10}$$

This expression is indeed timelike, but is not necessarily future-directed: Eqn. (6) is a solution of the Klein-Gordon equation whether k_a is future- or past-directed. Thus, we have not succeeded in interpreting a solution of the Klein-Gordon equation in terms of a "probability density for finding the particle."

We next consider the situation with regard to the initial value problem. Since the Schrödinger equation is first order in time derivatives, a solution of that equation is uniquely specified by giving $\psi'(x)$ at some initial time, say $t = 0$. The Klein-Gordon equation, on the other hand, is second order in time derivatives. Hence, to specify a solution, one must give both ϕ and $\partial\phi/\partial t$ at the initial time $t = 0$. This radical change in the structure of the initial data is clearly a consequence of our having "squared" Eqn. (4). It is still another indication that the transition to the relativistic case is not just a trivial application of the Schrödinger prescription.

Finally, let's look briefly at the structure of the space of solutions of the Klein-Gordon equation. In the non-relativistic case, the space of solutions of the

Schrödinger equation is a Hilbert space: it's obviously a complex vector space, and we define the norm of the state ψ by:

$$\|\psi\|^2 = \int_{t=\text{const.}} \psi\psi^* \mathrm{d}V \tag{11}$$

That the real number (11) is independent of the $t = $ const. surface over which the integral is performed is a consequence of Eqn. (8) (assuming, as one always does, that ψ falls off sufficiently quickly at infinity.) One might therefore be tempted to try to define the norm of a solution of the Klein-Gordon equation as an integral of J_a

$$\int_S J_a \mathrm{d}s^a \tag{12}$$

over a spacelike 3-plane S. But it is clear from (10) that the expression (12) will not in general be nonnegative. Thus, the most obvious way to make a Hilbert space out of solutions of the Klein-Gordon equation fails. This, of course, is rather embarrassing, for we are used to doing quantum theory in a Hilbert space, with Hermitian operators representing observables, etc.

To summarize, a simple "relativization" of the Schrödinger equations leads to a number of maladies.

2. Hilbert Space and Operators

The collection of states of a quantum system, together with certain of the structure naturally induced on this collection, is described by a mathematical object known as a Hilbert space. We recall the basic definitions.

A *Hilbert space* consists, first of all, of a set H. Secondly, H has the structure of an Abelian group. That is to say, given any two elements, ξ and η, of H, there is associated a third element of H, written $\xi + \eta$, this operation subject to the following conditions:

H1. For $\xi, \eta \in H$, $\xi + \eta = \eta + \xi$.

H2. For $\xi, \eta, \phi \in H$, $(\xi + \eta) + \phi = \xi + (\eta + \phi)$.

H3. There is an element of H, written "0", with the following property: for each $\xi \in H$, $\xi + 0 = \xi$.

H4. If $\xi \in H$, there exists an element of H, written "$-\xi$", with the following property: $\xi + (-\xi) = 0$.

Furthermore, H has the structure of a complex vector space. That is to say, with each complex number α and each element ξ of H there is associated an element of H, written $\alpha\xi$, this operation subject to the following conditions:

H5. For $\xi, \eta \in H$, $\alpha \in \mathbb{C}$, $\alpha(\xi + \eta) = \alpha\xi + \alpha\eta$.

H6. For $\xi \in H$, $\alpha, \beta \in \mathbb{C}$, $(\alpha + \beta)\xi = \alpha\xi + \beta\xi$ and $(\alpha\beta)\xi = \alpha(\beta\xi)$.

H7. For $\xi \in H$, $1\xi = \xi$.

There is, in addition, a positive-definite inner product defined on H. That is to say, with any two elements, ξ and η, of H there is associated a complex number, written (ξ, η), this operation subject to the following conditions:

H8. For $\xi, \eta, \phi \in H$, $\alpha \in \mathbb{C}$, $(\alpha\xi + \eta, \phi) = \alpha(\xi, \phi) + (\eta, \phi)$.

H9. For $\xi, \eta \in H$, $\overline{(\xi, \eta)} = (\eta, \xi)$.

H10. For $\xi \in H$, with $\xi \neq 0$, $(\xi, \xi) > 0$. (That (ξ, ξ) is real follows from **H9**.)

We sometimes write $\|\xi\|$ for $\sqrt{(\xi, \xi)}$. Finally, we require that this structure have a property called completeness. A sequence, ξ_i ($i \in 1, 2, \ldots$), of elements of H is called a *Cauchy sequence* if, for every number $\epsilon > 0$, there is a number N such that $\|\xi_i - \xi_j\| < \epsilon$ whenever i and j are greater than N. A sequence is said to *converge* to $\xi \in H$ if $\|\xi - \xi_j\| \to 0$ as $i \to \infty$. H is said to be *complete* if every Cauchy sequence converges to an element of H.

H11. H is complete.

There are, of course, hundreds of elementary properties of Hilbert spaces which follow directly from these eleven axioms.

A (linear) operator on a Hilbert space H is a rule A which assigns to each element ξ of H another element of H, written $A\xi$, this operation subject to the following condition:

O1. $\xi, \eta \in H$, $\alpha \in \mathbb{C}$, $A(\alpha\xi + \eta) = \alpha A\xi + A\eta$.

We shall discuss the various properties and types of operators when they arise.

There is a fundamental difficulty which arises when one attempts to use this mathematical apparatus in physics. The "collection of quantum states" which arises naturally in a physical problem normally satisfies **H1–H10**. (This is usually easy to show in each case.) The problem is with **H11**. The most obvious collection of states often fails to satisfy the completeness condition. As one wants a Hilbert space, he normally corrects this deficiency by completing the space, that is, by including additional elements so that all Cauchy sequences will have something to converge to. (There is a well-defined mathematical procedure for constructing, from a space which satisfies **H1–H10**, a Hilbert space.) The unpleasant consequence of being forced to introduce these additional states is that the natural operators of the problem, which were defined on the original collection of states, cannot be defined in any reasonable way on the entire Hilbert space. Thus, they are not operators at all as we have defined them, for they only operate on a subset of the Hilbert space. Fortunately, this subset is dense. (A subset D of a Hilbert space H is said to be dense if, for every element ξ of H, there is a sequence consisting of elements of D which converges to ξ.) Some very unaesthetic mathematical techniques have been devised for dealing with such situations. (See Von Neumann's book on Mathematical Foundations of Quantum Mechanics.)

This problem is not confined to quantum field theory. It occurs already in Schrödinger theory. For example, the collection of smooth solutions of the Schrödinger equation for which the integral (11) converges satisfy **H1–H10**, but not **H11**. To complete this space, we have to introduce "solutions" which are, for example, discontinuous. How does one apply the Schrödinger momentum operator to such a wave function?

3. Positive-Frequency Solutions of the Klein-Gordon Equation

We represent solutions of the Klein-Gordon equation as linear combinations of plane-wave solutions (Eqn. (6)):

$$\phi(x) = \int_{M_\mu} f(k_a) e^{ik_a x^a} \, dV_\mu \tag{13}$$

Of course, we wish to include in the integral (13) only plane-waves which satisfy the Klein-Gordon equation, i.e., only plane waves whose k_a satisfy the normalization condition (7). The four-dimensional (real) vector space of constant vector fields in Minkowski space-time is called *momentum space*. The collection of all vectors k_a in momentum space which satisfy Eqn. (7) consists of two hyperbolas (except in the case $\mu = 0$, in which case the hyperbolas degenerate to the two null cones through the origin). This collection is called the *mass shell* (associated with μ), M_μ (Fig. 3.1). Thus the function f in (13) is defined only on the mass shell, and the integral is to be carried out over the mass shell. It is convenient, furthermore, to distinguish the *future mass shell* M_μ^+ (consisting of future-directed vectors which satisfy (7)) and the *past mass shell* M_μ^- (consisting of past-directed vectors which satisfy (7)), so $M_\mu = M_\mu^+ \cup M_\mu^-$.

Eqn. (13) immediately suggests two questions: i) What are the necessary and sufficient conditions on the complex-valued function f on M_μ in order that the integral (13) exist for every x^a, and in order that the resulting $\phi(x)$ be

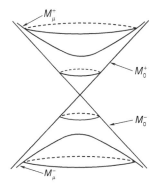

Figure 3.1: The mass shell in momentum space.

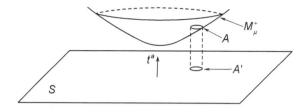

Figure 3.2:   The volume element on the mass shell.

smooth and satisfy the Klein-Gordon equation? ii) What are the necessary and sufficient conditions on a solution $\phi(x)$ of the Klein-Gordon equation in order that it can be expressed in the form (13) for some f? These, of course, are questions in the theory of Fourier analysis. It suffices for our purposes, however, to remark that the required conditions are of a very general character (that functions not be too discontinuous, and that, asymptotically, they go to zero sufficiently quickly). The point is that all the serious things we shall do with the Klein-Gordon equation will be in momentum space. We shall use Minkowski Space and $\phi(x)$ essentially only to motivate definitions and constructions on the f's in momentum space.

One question regarding (13) which must be answered is what is the volume element dV_μ we are using on the mass shell. Of course, it doesn't make any real difference, for a change in the choice of volume element would merely result in a suitable readjustment of the f's. Our choice can therefore be dictated by convenience. We require that our volume element be invariant under Lorentz transformations on momentum space (note that these leave the mass shell invariant), and that it be applicable also in the case $\mu = 0$. It is easy to state an appropriate volume element in geometrical terms. Let $\mu > 0$. Then the mass shell is a spacelike 3-surface in momentum space, in which there is a metric, so a metric is induced on M_μ. A metric on this 3-manifold defines a volume element $d\tilde{V}_\mu$. This $d\tilde{V}_\mu$ is clearly Lorentz-invariant, but, unfortunately, it approaches zero as $\mu \to 0$. To correct this, we define

$$dV_\mu = \mu^{-1}d\tilde{V}_\mu \qquad (14)$$

which is easily verified to he nonzero also on the null cone. In more conventional terms, our volume-element can be described as follows. Choose a unit time-like vector t^a in momentum space, and let S be the spacelike 3-plane perpendicular to t^a. Then any small patch A on M_μ, located at the point k_a, can be projected along t^a to give a corresponding patch A' on S (see Fig. 3.2.) Let dV'_μ be the volume of A' on S (using the usual volume element in the 3-space S). Our volume element on M_μ is then given by the following expression:

$$dV_\mu = dV'_\mu |t^a k_a|^{-1} \qquad (15)$$

The existence of a limit as $\mu \to 0$ is clear from (15), but Lorentz-invariance (independence of the choice of t^a) is not.

Is there any "gauge" in f? Given a solution $\phi(x)$ of the Klein-Gordon equation, is f uniquely determined by (13)? The only arbitrary choice which was made in writing (13) was the choice of an origin: "x^a" refers to the position vector of a point in Minkowski space with respect to a fixed origin. We are thus led to consider the behavior of the f's under origin changes. Let O and O' be two origins, and let v_a be the position vector of O' relative to O (see Fig. 3.3.) Let the position vectors of a point p in Minkowski space with respect to 0 and

Figure 3.3: Illustration of vectors in Minkowski space.

O' be x^a and x'^a, respectively, whence

$$x'^a = x^a - v^a \tag{16}$$

Then, if f and f' are the Fourier transforms of ϕ with respect to the origins O and O', respectively, we have

$$\phi(p) = \int_{M_\mu} f e^{ik_a x^a} \, dV_\mu = \int_{M_\mu} f' e^{ik_a x'^a} \, dV_\mu \tag{17}$$

Clearly, we must have

$$f'(k) = f(k) e^{ik_a v^a} \tag{18}$$

Thus, when we consider states as represented by functions on the mass shell, it is necessary to check that conclusions are unchanged if (18) is applied simultaneously to all such functions.

Now look again at Eqn. (3). It says, in particular, that the energy-momentum vector is future-directed. This same feature shows up in the right side of Eqn. (4) by the plus sign. If this sign were replaced by a minus, we would be dealing with a past-directed energy-momentum vector. The trick we used to obtain Eqn. (5) from (4) amounted to admitting also past-directed energy-momenta. It is clear now how Eqn. (4) itself can be carried over into a well-defined and fully relativistic condition on ϕ. We merely require that the f of Eqn. (13) vanish on M_μ^-. We call the corresponding solutions of the Klein-Gordon equation *positive-frequency* (or positive-energy) solutions. Defining negative-frequency solutions analogously, it is clear that every solution of the Klein-Gordon equation (more precisely, every solution which can be Fourier-analyzed) can be written uniquely as the sum of a positive-frequency and a negative-frequency solution. The positive-frequency solutions (resp., negative-frequency solutions) form a vector subspace of the vector space of all (Fourier analyzable) solutions.

To summarize, we are led to take as the "wave function of a free relativistic (spin zero) particle" a positive-frequency solution of the Klein-Gordon equation.

To what extent does this additional condition take care of the difficulties we discovered in Sect. 1?

Consider first the current, Eqn. (9). We saw before, from the example of a negative-frequency plane wave, that in general J is neither future-directed or even timelike. Is it true that J is future-directed timelike for a positive-frequency solution of the Klein-Gordon equation? Unfortunately, the answer is no. Consider a linear combination of two positive-frequency plane waves:

$$\phi = e^{ik_a x^a} + \alpha e^{ik'_a x^a} \tag{19}$$

That is, α is a complex constant, and k_a and k'_a are future-directed constant vectors satisfying (7). (Strictly speaking, this example is not applicable, for (19) cannot be Fourier analyzed. It is not difficult, however, to appropriately smear (19) over the future mass shell to obtain an example without this deficiency.) Substituting (19) into (9), we obtain:

$$
\begin{aligned}
J_a = & \frac{1}{2} k_a \left[2 + \alpha e^{i(k'_b - k_b)x^b} + \alpha^* e^{i(k_b - k'_b)x^b} \right] \\
& + \frac{1}{2} k'_a \left[2\alpha\alpha^* + \alpha e^{i(k'_b - k_b)x^b} + \alpha^* e^{i(k_b - k'_b)x^b} \right]
\end{aligned}
\tag{20}
$$

Clearly, one can choose α, k_a, and k'_a so that this J_a is not timelike in certain regions. Thus, even the assumption of positive-frequency solutions does not resolve the difficulty associated with not having a simple probabilistic interpretation for our wavefunction ϕ: we still cannot think of $J_a t^a$ (with t^a unit, future-directed, timelike) as representing a probability density for finding the particle. The resolution of this problem must await our introduction of a position operator.

Note from Eqn. (20) that J_a is trying very hard to be timelike and future-directed in the positive-frequency case: it is only the cross terms between the two plane waves which destroys this property. This observation suggests that, in the positive-frequency case, the integral of J_a over a spacelike 3-plane might be positive. In order to check on this possibility, we want to rewrite the integral of J_a in terms of the corresponding function f on M_μ. It will be more illuminating to do this for the general solution ϕ of the Klein-Gordon equation, i.e., not assuming, for the time being, that ϕ is positive-frequency. Substituting (13) into (9):

$$
\begin{aligned}
J_a = & \int_{M_\mu} dV_\mu \int_{M_\mu} dV'_\mu \frac{1}{2} k'_a \times \\
& \times \left[f^*(k) f(k') e^{i(k'_b - k_b)x^b} + f(k) f^*(k') e^{i(k_b - k'_b)x^b} \right]
\end{aligned}
\tag{21}
$$

We now let S be a spacelike 3-plane through the origin, and let t^a be the unit, future-directed normal to S. Then

$$
\begin{aligned}
\int_S J_a t^a \, dS = & \int_{M_\mu} dV_\mu \int_{M_\mu} dV'_\mu \frac{1}{2} k'_a t^a \left[f^*(k) f(k') \int_S e^{i(k'_b - k_b)x^b} \, dS \right. \\
& + \left. f(k) f^*(k') \int_S e^{i(k_b - k'_b)x^b} dS \right]
\end{aligned}
\tag{22}
$$

But, from the theory of Fourier analysis

$$\int_S e^{i(k_b' - k_b)x^b}\, \mathrm{d}S = (2\pi)^3 |t^a k_a|^{-1} \delta(k - k')$$

(23)

and so (22) becomes

$$\int_S J_a t^a\, \mathrm{d}S = (2\pi)^3 \left[\int_{M_\mu^+} ff^*\, \mathrm{d}V_\mu - \int_{M_\mu^-} ff^*\, \mathrm{d}V_\mu \right]$$

(24)

In particular, if f vanishes on M_μ^-, the left side of (24) must be positive, or vanishes if and only if ϕ vanishes. (The actual form of (24) was rather suggested by Eqn. (15). If we had omitted the absolute-value sign on the right (and what better thing could be done with an absolute-value sign?), the "volume element" on M_μ^- would have been negative.) This calculation was not done merely for idle curiosity; the right side of (24) will be important later.

We saw before that the initial-value problem for the Klein-Gordon equation is as follows: one must specify ϕ and $t^a \nabla_a \phi$ on an initial spacelike 3-plane. How does the initial-value problem go for positive-frequency solutions of the Klein-Gordon equation? In fact, we only have to specify ϕ as initial data in this case. To see this, suppose we know the value of the integral

$$\phi(x) = \int_{M_\mu^+} f(k) e^{ik_a x^a}\, \mathrm{d}V_\mu$$

(25)

for every x^a which is perpendicular to a unit timelike vector t^a at the origin (i.e., on the spacelike 3-plane perpendicular to t^a, through the origin). The integral (25) can certainly be expressed as a Fourier integral over S (t^a sets up a one-to-one correspondence between M_μ^+ and S). But then, by taking a Fourier transform, we can determine f on M_μ^+. Thus, we know ϕ throughout Minkowski space. That is to say, when we properly interpret (4), we obtain a "Schrödinger-type" initial-value problem. If we ignore questions of smoothness and convergence of Fourier integrals, the situation can be roughly summarized as follows:

1. There is a one-to-one correspondence between: i) solutions of the Klein-Gordon equation, ii) complex-valued functions f on M_μ, and iii) values of ϕ and $t^a \nabla_a \phi$ on a spacelike 3-plane.

2. There is a one-to-one correspondence between: i) positive-frequency solutions of the Klein-Gordon equation, ii) complex-valued functions on M_μ^+, and iii) values of ϕ on a spacelike 3-plane.

4. Constructing Hilbert Spaces and Operators

There is a general and extremely useful technique for obtaining a Hilbert space along with a collection of operators on it. It is essentially this technique which is used, for example, in treating the Schrödinger and Klein-Gordon equations. It is convenient, therefore, to describe this construction, once and for all, in a general case. Special cases can then be treated as they arise.

The fundamental object we need is some n-dimensional manifold M on which there is specified a smooth, nowhere-vanishing volume-element dV. In differential-geometric terms, this means that we have a smooth, nowhere-vanishing, totally skew tensor field $\epsilon_{a_1 \cdots a_n}$ on M. Our Hilbert space, and operators, are now defined in terms of certain fields on M.

We first define the Hilbert space. Consider the collection H of all complex-valued, measurable, square-integrable functions f on M. This H is certainly a complex vector apace. We introduce a norm on H:

$$\|f\|^2 = \int_M f f^* \, dV \qquad (26)$$

It is known that this H thus becomes a Hilbert space. (Actually, we have been a little sloppy here. One should, more properly, define an equivalence relation on H: two functions are equivalent if they differ only on a subset (of M) of measure zero. It is the equivalence classes which actually form a Hilbert space. For example, the function f which vanishes everywhere on M except one point, where it is one, is measurable and square-integrable. Its norm, (26), is zero, although this f is not the zero element of H. It is, however, in the zero equivalence class, for it differs from the zero function only on a set (namely, one point) of measure zero.) This is a special case of a more general theorem: the collection of all complex-valued, measurable, square-integrable functions (more precisely, the collection of equivalence classes as above) on a complete measure space form a Hilbert space.

We now introduce some operators. Let v^a be any smooth (complex) contravariant vector field, and v any smooth (complex) scalar field on M. Then with each smooth, complex-valued function f on M we may associate the function

$$Vf = v^a \nabla_a f + vf \qquad (27)$$

13

where ∇_a denotes the gradient on M. To what extent does (27) define an operator on H? Unfortunately, (27) is not applicable to every element of H, for two reasons: i) a function f could be measurable and square-integrable (i.e., an element of H), but not differentiable. Then the gradient operation in (27) would not be defined. ii) an element f of H could even be smooth, but could have the property that, although f itself is square-integrable, the function (27) is not. However, there is a large class of elements of H on which (27) is defined and results in an element of H. Such a class, for example, is the collection of all functions F which are smooth and have compact support. (Such a function is automatically square-integrable and measurable.) This class is, in fact, dense in H. Clearly, (27) is linear whenever it is defined. Thus, we can call (27) an "operator on H", in the sense that we have agreed to abuse that term.

We agree to call an operator *Hermitian* if, whenever Vf and Vg are defined, $(Vf, g) = (f, Vg)$. What are the necessary and sufficient conditions that (27) be Hermitian? Let f and g be smooth functions on M, of compact support. Then:

$$
\begin{aligned}
(Vf, g) &= \int_M (v^a \nabla_a f + vf) g^* \, \mathrm{d}V \\
&= \int_M [-f v^a \nabla_a g^* + f g^*(-\nabla_a v^a + v)] \, \mathrm{d}V
\end{aligned}
\tag{28}
$$

where we have done an integration by parts (throwing away a surface term by the compact supports). Eqn. (28) is clearly equal to

$$
(f, Vg) = \int_M [f v^{*a} \nabla_a g^* + f v^* g^*] \, \mathrm{d}V
\tag{29}
$$

for every f and g when and only when:

$$
v^{*a} = -v^a \qquad v - v^* = \nabla_a v^a
\tag{30}
$$

These, then, are the necessary and sufficient conditions that V be Hermitian. One further remark is required with regard to what the divergence in the first Eqn. (30) is supposed to mean. (We don't have a metric, or a covariant derivative, defined on M.) It is well-known that the divergence of a contravariant vector field can be defined on a manifold with a volume-element $\epsilon_{a_1 \cdots a_n}$. This can be done, for example, using either exterior derivatives or Lie derivatives. For instance, using Lie derivatives we define "$\nabla_a v^a$" by:

$$
\mathscr{L}_{v^m} \epsilon_{a_1 \cdots a_n} = (\nabla_a v^a) \epsilon_{a_1 \cdots a_n}
\tag{31}
$$

(Note that, since the left side is totally skew, it must be *some* multiple of $\epsilon_{a_1 \cdots a_n}$.)

Finally, we work out the commutator of two of our operators; $V = (v^a, v)$ and $W = (w^a, w)$. If f is a smooth function of compact support, we have:

$$
\begin{aligned}
[V, W]f &= (v^a \nabla_a + v)(w^b \nabla_b + w)f - (w^b \nabla_b + w)(v^a \nabla_a + v)f \\
&= (v^b \nabla_b w^a - w^b \nabla_b v^a) \nabla_a f + (v^a \nabla_a w - w^a \nabla_a v)f
\end{aligned}
\tag{32}
$$

Note that the commutator is again an operator of the form we have been discussing, (27). Note furthermore that the vector part of the commutator is the Lie bracket of the vector fields appearing in V and W.

To summarize, with any n-manifold M on which there is given a smooth, nowhere-vanishing volume element we associate a Hilbert space H along with a collection of operators on H. The commutator of two operators in this collection is again an operator in the collection.

5. Hilbert Space and Operators for the Klein-Gordon Equation

We now complete our description of the quantum theory of a free, relativistic, spin-zero particle.

For our Hilbert space we take, as suggested by Sec. 3, the collection of all complex-valued, measurable, square-integrable functions on the future mass shell, M_μ^+. In order to obtain position, momentum, energy, etc. operators, we use the scheme described in Sec. 4. That is, we look for vector and scalar fields on M_μ^+.

We first consider momentum operators. Let p^a be any constant vector field in Minkowski apace, and ϕ any positive-frequency solution of the Klein-Gordon equation. Then, clearly,

$$\frac{\hbar}{i} p^a \nabla_a \phi \tag{33}$$

is also a positive-frequency solution of the Klein-Gordon equation. In terms of the corresponding functions on M_μ^+, (33) takes the form

$$f \to (\hbar p^a k_a) f \tag{34}$$

That is to say, we multiply f by the real function $(\hbar p^a k_a)$ on M_μ^+. Thus, for each constant vector field p^a, we have an operator, $P(p^a)$, on our Hilbert space H. Since the multiplying function in (34) is real, the operators $P(p^a)$ are all Hermitian. (See (30).) We now interpret these operators. Choose a constant, unit, future-directed timelike vector field t^a in Minkowski space (a preferred "state of rest"). Then $P(t^a)$ is the "energy" operator, and $P(p^a)$, with p^a unit and perpendicular to t^a, is the "component of momentum in the p^a-direction" operator.

The position operators are more complicated. Not only do they depend on more objects in Minkowski space (rather than just a single p^a as in the momentum case), but also they require us to take derivatives in the mass shell. To obtain a position operator, we need the following information: a choice of origin O in Minkowski space, a constant, unit, future-directed timelike vector field t^a in Minkowski space, and a constant unit vector field q^a which is perpendicular to t^a. (Roughly speaking, O and t^a define a spacelike 3-plane — the "instant" at which the operator is to be applied — q^a defines "which position coordinate we're operating with", and O tells us what the origin of this position coordinate is.) Now, q^a is a vector in momentum space, and therefore defines a constant vector field in momentum space, which we also write as q^a. One is tempted to take the derivative of f along this vector field. But this will not work, for q^a is not tangent to the mass shell, whereas f is only defined on the mass shell. To correct this deficiency, we project q^a into the mass shell — that is, we add to q^a that multiple of t^a which results in a vector field lying in M_μ^+:

$$-\frac{1}{i}\left[q^a - t^a(t^b k_b)^{-1}(q^c k_c)\right] \tag{35}$$

We now have a vector field on M_μ^+, and therefore an operator on our Hilbert space H. But are those operators Hermitian? From (30), we see that this question reduces to the question of whether the divergence of (35) vanishes or not. Unfortunately, we obtain for this divergence

$$-\frac{1}{i}\left(g^{ab} - \mu^{-2}k^a k^b\right)\partial_a\left[q^b - t^b(t^c k_c)^{-1}(q^d k_d)\right] = -\frac{1}{i}(q^a k_a)(t^b k_b)^{-2} \tag{36}$$

where we have denoted the derivative in momentum space by ∂_a. To obtain a Hermitian operator, we take the Hermitian part of the operator represented by (35):

$$f \to -\frac{1}{i}\left[q^a - t^a(t^b k_b)^{-1}(q^c k_c)\right]\partial_a f - \frac{1}{2i}(q^a k_a)(t^b k_b)^{-2} \tag{37}$$

In (37), f is to be the function on M_μ^+ obtained using O as the origin (see (18).) (Why is there no $\hbar$ in (37)? We should, perhaps, have called k-space "wave number and frequency space" rather than "momentum space".) We shall write the operator (37) $X(O, t^a, q^a)$. For any value of its arguments, X is a Hermitian operator on H. (It is strange — and perhaps rather unpleasant — that the position and momentum operators are so different from each other.)

We now have a lot of operators, and so we can ask for their commutators. This is easily done by substituting into our explicit formula, Eqn. (32). The result is the standard formulae:

$$\begin{aligned}
&\left[P(p^a), P(p'^a)\right] = 0\\
&\left[X(O, t^a, q^a), X(O, t^a, q'^a)\right] = 0\\
&\left[P(p^a), X(O, t^a, q^a)\right] = -i\hbar(p^a q_a)\\
&(p^a t_a) = 0
\end{aligned} \tag{38}$$

The next thing one normally does with operators (in the Heisenberg representation, which is the one we're using) is to work out their time derivatives. For the momentum operators, this is easy, for no notion of a "time" was used to define $P(p^a)$). Thus, whatever reasonable thing one wants to mean by a " · ", we have:

$$\dot{P}(p^a) = 0 \tag{39}$$

This, of course, is what we expect for the momentum operator on a free particle. For the position operators, on the other hand, we have an interesting notion of time-derivative. We want to compare $X(O, t^a, q^a)$ with "the same position operator at a slightly later time". This "at a slightly later time" is expressed by slightly displacing O in the t^a-direction. Thus, we are led to define:

$$\dot{X}(O, t^a, q^a) = \lim_{\epsilon \to 0} \frac{1}{\epsilon} [X(O', t^a, q^a) - X(O, t^a, q^a)] \tag{40}$$

where O' is defined by the property that its position vector relative to O is ϵt^a. It is straightforward to check, with this definition, that

$$\dot{X}(O, t^a, q^a)f = -(q^a k_a)(t^b k_b)^{-1} f \tag{41}$$

which, of course, is what we expected. Note that a number of statements about how $X(O, t^a, q^a)$ depends on its arguments follow directly from Eqn. (41).

Finally, one would like to ask about the eigenvectors and eigenvalues of our operators. It is clear from Eqn. (34) that the only candidate for an eigenfunction of $P(p^a)$ would be a δ-function on M_μ^+. Of course, a δ-function is not a function, and hence not an element of H (we cannot enlarge H to include such functions, if we want to keep a Hilbert space, for a δ-function should not be square-integrable.) It is convenient to have the idea, however, that if $P(p^a)$ had eigenfunctions, they would be plane-waves. We next ask for eigenfunctions of $X(O, t^a, q^a)$. We look for the wave function of a "particle localized at the origin", that is we look for an f such that $X(O, t^a, q^a)f = 0$ for *every* q^a which is perpendicular to t^a (t^a and O fixed). That is, from (37), we require that

$$[q^a - t^a(t^b k_b)^{-1}(q^c k_c)] \, \partial_a f - \frac{1}{2}(q^a k_a)(t^b k_b)^{-2} f = 0 \tag{42}$$

for every such q^a. The solution to (42) is:

$$f = \text{const.} \, (t^a k_a)^{1/2} \tag{43}$$

The first remark concerning (43) is that it is not square-integrable, and hence does not represent an element of H. This does not stop us, however, from substituting (43) into (13) to obtain a function ϕ on Minkowski space. The resulting ϕ (the explicit formula is not very enlightening — it involves Hankel functions) is not a δ-function at O. In fact, this ϕ is "spread out" around O to distances of the order of μ^{-1} — the Compton wavelength of our particle. Thus, our picture is that a relativistic particle cannot be confined to distances much smaller than its Compton wavelength.

6. The Direct Sum of Hilbert Spaces

Associated with any countable sequence, $H', H'', H''', \ldots$, of Hilbert spaces there is another Hilbert space, written $H' \oplus H'' \oplus H''' \oplus \ldots$, and called the *direct sum* of $H', H'', H''', \ldots$. We shall give the definition of the direct sum and a few of its elementary properties.

Consider the collection of all sequences

$$(\xi', \xi'', \xi''', \ldots) \tag{44}$$

consisting of one element (ξ') of H', one element (ξ'') of H'', etc., for which the sum

$$\|\xi'\|^2 + \|\xi''\|^2 + \|\xi''\|^2 + \ldots \tag{45}$$

converges. This collection is the underlying point set of the direct sum. To obtain a Hilbert space, we must define addition, scalar multiplication, and an inner product, and verify **H1–H11**.

The sum of two sequences (44) is defined by adding them "component-wise":

$$(\xi', \xi'', \xi''', \ldots) + (\eta', \eta'', \eta''', \ldots) = (\xi' + \eta', \xi'' + \eta'', \xi''' + \eta''', \ldots) \tag{46}$$

We must verify that, if the addends satisfy (45), then so does the sum. This follows immediately from the inequality:

$$\begin{aligned}
\|\xi' + \eta'\|^2 &= \|\xi'\|^2 + (\xi', \eta') + (\eta', \xi') + \|\eta'\|^2 \\
&\leq \|\xi'\|^2 + 2\|\xi'\|\|\eta'\| + \|\eta'\|^2 \\
&\leq 2\|\xi'\|^2 + 2\|\eta'\|^2
\end{aligned} \tag{47}$$

The product of a sequence (44) and a complex number α is defined by:

$$\alpha(\xi', \xi'', \xi''', \ldots) = (\alpha\xi', \alpha\xi'', \alpha\xi''', \ldots) \tag{48}$$

That the right side of (48) satisfies (45) follows from the fact that

$$\|\alpha\xi'\| = |\alpha|\|\xi'\| \tag{49}$$

We have now defined addition and scalar multiplication. That these two operations satisfy **H1–H7**, i.e., that we have a complex vector space, is trivial.

We define the inner product between two sequences (44) to be the complex number

$$((\xi', \xi'', \xi''', \ldots), (\eta', \eta'', \eta''', \ldots)) = (\xi', \eta') + (\xi'', \eta'') + (\xi'', \eta'') + \ldots \quad (50)$$

The indicated sum of complex numbers on the right of (50) converges if (in fart, converges absolutely if and only if) the sum of the absolute values converges. Thus, the absolute convergence of the right side of (50) follows from the fact that

$$|(\xi', \eta')| \leq \|\xi'\| \leq \frac{1}{2}\|\xi'\|^2 + \frac{1}{2}\|\eta'\|^2 \quad (51)$$

We now have a complex vector space in which there is defined an inner product. (Note incidentally, that the norm is given by (45).) The verification of **H8**, **H9**, and **H10** is easy.

Thus, as usual, the only difficult part is to check **H11**. Consider a Cauchy sequence of sequences (44). That is to say, we have a countable collection of such sequences,

$$\phi_1 = (\xi_1', \xi_1'', \xi_1''', \ldots)$$
$$\phi_2 = (\xi_2', \xi_2'', \xi_2''', \ldots)$$
$$\phi_3 = (\xi_3', \xi_3'', \xi_3''', \ldots) \quad (52)$$
$$\vdots$$

with the following property: for each real $\epsilon > 0$ there is a number N such that

$$\|\phi_i - \phi_j\|^2 = \|\xi_i' - \xi_j'\|^2 + \|\xi_i'' - \xi_j''\|^2 + \cdots \leq \epsilon \quad (53)$$

whenever $i, j \geq N$. We must show that the sequence of elements (52) of the direct sum converge to some element of the direct sum. First note that (53) implies

$$\|\xi_i' - \xi_j'\|^2 \leq \epsilon, \quad \|\xi_i'' - \xi_j''\|^2 \leq \epsilon, \quad \ldots \quad (54)$$

That is to say, the first "column" of (52) is a Cauchy sequence in H', the second column a Cauchy sequence in H'', etc. Since $H', H'', \ldots$ are Hilbert spaces, these Cauchy sequences converge, say, to $\xi' \in H'$, $\xi'' \in H''$, etc. Form

$$\phi = (\xi', \xi'', \xi''', \ldots) \quad (55)$$

We must show that the ϕ_i converge to ϕ, and that is ϕ is an element of the direct sum (i.e., that (45) converges for ϕ). Fix $\epsilon \geq 0$ and choose i such that $\|\phi_i - \phi_j\|^2 \leq \epsilon$ whenever $j > i$. Then, for each positive integer n,

$$\|\xi_i' - \xi_j'\|^2 + \|\xi_i'' - \xi_j''\|^2 + \ldots + \|\xi_i^{(n)} - \xi_j^{(n)}\|^2 \leq \epsilon \quad (56)$$

Taking the limit of (56) as $j \to \infty$, we obtain

$$\|\xi_i' - \xi'\|^2 + \|\xi_i'' - \xi''\|^2 + \ldots + \|\xi_i^{(n)} - \xi^{(n)}\|^2 \leq \epsilon \quad (57)$$

but n is arbitrary, and so, taking the limit of (57) as $n \to \infty$,

$$\|\phi_i - \phi\|^2 = \|\xi_i' - \xi'\|^2 + \|\xi_i'' - \xi''\|^2 + \ldots \leq \epsilon \tag{58}$$

That is to say, the ϕ_i converge to ϕ. Finally, the fact that ϕ is actually an element of the direct sum, i.e., the fact that

$$\|\xi'\|^2 + \|\xi''\|^2 + \|\xi''\|^2 + \ldots \tag{59}$$

converges, follows immediately by substituting

$$\|\xi'\|^2 \leq 2\|\xi_i'\|^2 + 2\|\xi_i' - \xi'\|^2 \tag{60}$$

(and the corresponding expressions with more primes) into (59), and using (58) and the fact that ϕ_i is an element of the direct sum. Thus, the direct sum is complete.

To summarize, we have shown how to construct a Hilbert space from a countable sequence of Hilbert spaces. Note, incidentally, that the direct sum is essentially independent of the order in which the Hilbert spaces are taken. More precisely, the direct sum obtained by taking $H', H'', \ldots$ in one order is naturally isomorphic to the direct sum obtained by taking these spaces in any other order.

Finally, we discuss certain operators on the direct sum. Consider a sequence of operators: A' acting on H', A'' acting on H'', etc. Then with each element (44) of the direct sum we may associate the sequence

$$(A'\xi', A''\xi'', A'''\xi''', \ldots) \tag{61}$$

Unfortunately, (61) may not be an element of the direct sum, for

$$\|A'\xi'\|^2 + \|A''\xi''\|^2 + \|A'''\xi'''\|^2 + \ldots \tag{62}$$

may fail to converge. However, (61) will produce an element of the direct sum when acting on a certain dense subset of the direct sum, namely, the set of sequences (44) which consist of zeros after a certain point. Is there any condition on the A's which will ensure that (61) will always be an element of the direct sum? An operator A on a Hilbert space H is said to be *bounded* if A is defined everywhere and, for some number a, $\|A\xi\| \leq a\|\xi\|$ for every $\xi \in H$. The smallest such a is called the *bound* of A, written $|A|$. (The norm on a Hilbert space induces on it a metric topology. Boundedness is equivalent to continuity in this topology.) It is clear that (62) converges for every element of the direct sum provided i) all the A's are bounded, and ii) the sequence of real numbers $|A'|, |A''|, \ldots$ is bounded.

7. The Completion of an Inner-Product Space

It is sometimes the case, when one wishes to construct a Hilbert space, that one finds a set on which addition, scalar multiplication, and an inner product are defined, subject to **H1–H10** — what we shall call an inner product space. One wants, however, to obtain a Hilbert space, i.e., something which also satisfies **H11**. There is a construction for obtaining a Hilbert space from an inner product space. Since this construction is in most textbooks, we merely indicate the general idea.

Let G be an inner product space. Denote by G' the collection of all Cauchy sequences in G'. If $\xi_i \in G', \eta_i \in G'$ $(i = 1, 2, \ldots)$, we write $\xi_i \approx \eta_i$ provided

$$\lim_{i \to \infty} \|\xi_i - \eta_i\| = 0 \tag{63}$$

One verifies that "$\approx$" is an equivalence relation. The collection of equivalence classes, denoted by $\overline{G}$, is to be made into a Hilbert space.

Consider two elements of $\overline{G}$, i.e., two equivalence classes, and let ξ_i and η_i be representatives. We define a new sequence, whose ith element is $\xi_i + \eta_i$. One verifies, using the fact that ξ_i and η_i, are Cauchy sequences, that this new sequence is also Cauchy. Furthermore, if ξ_i and η_i are replaced by equivalent Cauchy sequences, the sum becomes a Cauchy sequence which is equivalent to $\xi_i + \eta_i$. Thus, we have defined an operation of addition in $\overline{G}$. In addition, if ξ_i is a Cauchy sequence and α a complex number, $\alpha\xi_i$ is a Cauchy sequence whose equivalence class depends only on the equivalence class of ξ_i. We have thus defined an operation of scalar multiplication in $\overline{G}$. These two operations satisfy **H1–H7**.

If $\xi_i \in G'$, $\eta_i \in G'$, then

$$\lim_{i \to \infty} (\xi_i, \eta_i) \tag{64}$$

exists. Furthermore, this complex number is unchanged if ξ_i and η_i are replaced by equivalent Cauchy sequences. Thus, (64) defines an inner product on $\overline{G}$. One must now verify **H8**, **H9**, and **H10**, so that $\overline{G}$ becomes an inner product space. Finally (and this is the only hard part), one proves that $\overline{G}$ is complete, and so constitutes a Hilbert space. The Hilbert space $\overline{G}$ is called the *completion* of the inner product space G.

Note that G can be considered as a subspace of its completion; with each $\xi \in G$ associate the element of $\overline{G}$ (the equivalence class) containing the Cauchy sequence $\underline{\xi}_i = \xi$ of G. It is easily checked from the definition, in fact, that G is dense in $\overline{G}$. Suppose that G itself were already complete? Then every Cauchy sequence in G would converge, and, from (63), two Cauchy sequences would be equivalent if and only if they converged to the same thing. Thus, in this case $\overline{G}$ would be just another copy of G; the completion of a complete space is just that space again.

Finally, suppose that A is a bounded operator on the inner product space G. Then A can be extended to a bounded operator on $\overline{G}$. That is, there is a bounded operator on $\overline{G}$ which reduces to A on G considered as a subspace of $\overline{G}$. To prove this, let ξ_i be a Cauchy sequence in G. Then, since A is bounded,

$$\|A\xi_i - A\xi_j\| = \|A(\xi_i - \xi_j)\| \leq |A| \|\xi_i - \xi_j\| \tag{65}$$

whence $A\xi_i$ is a Cauchy sequence in G. Furthermore, if two Cauchy sequences satisfy (63), then

$$\lim_{i \to \infty} \|A\xi_i - A\eta_i\| \leq |A| \lim_{i \to \infty} \|\xi_i - \eta_i\| = 0 \tag{66}$$

That is to say, $A\xi_i$ is replaced by an equivalent Cauchy sequence when ξ_i is. Therefore, we can consider A as acting on elements of $\overline{G}$ to produce elements of $\overline{G}$. This action is clearly linear, and so we have an operator $\overline{A}$ defined on $\overline{G}$. Furthermore, this $\overline{A}$ is bounded, and in fact $|\overline{A}| = |A|$, for

$$\lim_{i \to \infty} \|A\xi_i\| \leq |A| \lim_{i \to \infty} \|\xi_i\| \tag{67}$$

for any Cauchy sequence in G.

8. The Complex-Conjugate Space of a Hilbert Space

Let H be a Hilbert space. We introduce the notion of the *complex-conjugate space* of H, written $\bar{H}$. As point-sets, $H = \bar{H}$. That is to say, with each element $\xi \in H$ there is associated an element of $\bar{H}$; this element will be written $\bar{\xi}$. Furthermore, we take as the group structure on $\bar{H}$ that is induced from H:

$$\bar{\xi} + \bar{\eta} = \overline{(\xi + \eta)} \tag{68}$$

In other words, the sum of two elements of $\bar{H}$ is defined by taking the sum (in H) of the corresponding elements of H, and taking the result back to $\bar{H}$. Scalar multiplication in H, on the other hand, is defined by the formula ($\mu \in \mathbb{C}, \bar{\xi} \in \bar{H}$):

$$\mu\bar{\xi} = \overline{(\bar{\mu}\xi)} \tag{69}$$

That is, to multiply an element of $\bar{H}$ by a complex number, one multiplies the corresponding element of H by the complex-conjugate of that number, and takes the result back to $\bar{H}$. (Note that a bar appears in two different senses in (69). A bar over a complex number denotes its complex-conjugate; a bar over an element of H denotes the corresponding element of $\bar{H}$.) Finally, the inner product on $\bar{H}$ is fixed by requiring that the transition from H to $\bar{H}$ preserve norms:

$$\|\bar{\xi}\| = \|\xi\| \tag{70}$$

It is obvious that this $\bar{H}$ thus becomes a Hilbert space.

Note that the complex-conjugate space of $\bar{H}$ is naturally isomorphic with H. We write $\bar{\bar{H}} = H$, and, for $\xi \in H$, $\bar{\bar{\xi}} = \xi$.

The reason for introducing $\bar{H}$ is that one frequently encounters mappings on H which are anti-linear ($T(\mu\xi + \eta) = \bar{\mu}T(\xi) + T(\eta)$) rather than linear ($T(\mu\xi + \eta) = \mu T(\xi) + T(\eta)$). Anti-linear mappings on H become linear mappings on $\bar{H}$, and it is easier to think about linear mappings than anti-linear ones. Consider, for example, the inner product on H, (ξ, η). This can be considered as a mapping $H \times H \to \mathbb{C}$, which is linear in the first H and anti-linear in the second. If, however, we consider the inner product as a mapping $H \times \bar{H} \to \mathbb{C}$, it becomes linear in both factors.

9. The Tensor Product of Hilbert Spaces

With any finite collection of Hilbert spaces, $H^\alpha, H^\beta, \ldots, H^\gamma$, there is associated another Hilbert space, called the *tensor product* of H^α, $H^\beta, \ldots, H^\gamma$, and written $H^\alpha \otimes H^\beta \otimes \cdots \otimes H^\gamma$. We shall define the tensor product and derive a few of its properties.

It is convenient to introduce an index notation. We attach a raised Greek index to a vector to indicate to which Hilbert space it belongs, e.g., $\xi^\alpha \in H^\alpha$, $\eta^\beta \in H^\beta$. The corresponding complex-conjugate spaces, $\overline{H^\alpha}, \overline{H^\beta}, \ldots, \overline{H^\gamma}$, will be written $\bar{H}_\alpha, \bar{H}_\beta, \ldots, \bar{H}_\gamma$. Membership in the complex-conjugate spaces will be indicated with a lowered Greek index. The element of $\bar{H}_\alpha$ which corresponds to $\xi^\alpha \in H^\alpha$ would be written $\bar{\xi}_\alpha$, while (69) would be written thus: $\mu\bar{\xi}_\alpha = \overline{(\bar{\mu}\,\xi^\alpha)}$. Finally, the operation of taking the inner product (which associates a complex number, linearly, with an element of H and an element of $\bar{H}$) is indicated by placing the two elements next to each other, e.g., $\xi^\alpha\bar{\eta}_\alpha$. Hence, $\|\xi^\alpha\|^2 = \xi^\alpha\bar{\xi}_\alpha$. The inner product operation looks (and is) similar to "contraction."

We now wish to define the tensor product. In order to avoid cumbersome strings of dots, we shall discuss the tensor product of just two Hilbert spaces, H^α and H^β. The tensor product of n Hilbert spaces is easily obtained by inserting dots at appropriate places in the discussion below.

Consider the collection of all formal expressions of the following type:

$$\xi^\alpha\eta^\beta + \ldots + \sigma^\alpha\tau^\beta \tag{71}$$

"Formal" here means that the pluses and juxtapositions of elements in (71) are not to be considered, for the moment, as well-defined operations. They are merely marks on the paper. We introduce, on the collection of all such formal sums, an equivalence relation: two formal sums will be considered equivalent if they can be obtained from each other by any combination of the following operations on such sums:

1. Permute, in any way, the terms of a formal sum.

2. Add to a formal sum, or delete, the following combination of terms: $(\mu\xi^\alpha)\eta^\beta + (-\xi^\alpha)(\mu^{-1}\eta^\beta)$.

3. Add to a formal sum, or delete, the following combination of terms: $(\xi^\alpha + \sigma^\alpha)\eta^\beta + (-\xi^\alpha)\eta^\beta + (-\sigma^\alpha)\eta^\beta$.

4. Add to a formal sum, or delete, the following combination of terms: $\xi^\alpha(\eta^\beta + \tau^\beta) + \xi^\alpha(-\eta^\beta) + \xi^\alpha(-\tau^\beta)$.

We denote the collection of equivalence classes by $F^{\alpha\beta}$. The idea is to introduce on this $F^{\alpha\beta}$ the structure of an inner-product space, and then take the completion to obtain the tensor product.

We add two formal sums by "stringing them together":

$$(\xi^\alpha\eta^\beta + \ldots + \sigma^\alpha\tau^\beta) + (\phi^\alpha\psi^\beta + \ldots + \lambda^\alpha\kappa^\beta) =$$
$$= \xi^\alpha\eta^\beta + \ldots + \sigma^\alpha\tau^\beta + \phi^\alpha\psi^\beta + \ldots + \lambda^\alpha\kappa^\beta \qquad (72)$$

The equivalence class of the formal sum on the right in (72) depends, of course, only on the equivalence classes of the two formal sums on the left, and so we have defined an operation of addition on $F^{\alpha\beta}$. Similarly, the product of a format sum and a complex number, defined by

$$\mu(\xi^\alpha\eta^\beta + \ldots + \sigma^\alpha\tau^\beta) = (\mu\xi^\alpha)\eta^\beta + \ldots + (\mu\sigma^\alpha)\tau^\beta \qquad (73)$$

induces an operation of scalar multiplication on $F^{\alpha\beta}$. (Note that we don't change the equivalence class of the right side of (73) by placing some or all of the μ's with the second vectors rather than the first.) Thus, we have on $F^{\alpha\beta}$ the structure of a complex vector space. So far, we have merely repeated the standard construction of the tensor product of two vector spaces.

We next wish to define an inner product, or, equivalently, a norm, on $F^{\alpha\beta}$. The norm of a formal sum, Eqn. (71), is defined by writing

$$(\xi^\alpha\eta^\beta + \cdots + \sigma^\alpha\tau^\beta)(\bar\xi_\alpha\bar\eta_\beta + \cdots + \bar\sigma_\alpha\bar\tau_\beta) \qquad (74)$$

and expanding using associativity. For example, the norm of a formal sum with just two terms would be given by the sum of the complex numbers on the right of:

$$(\xi^\alpha\eta^\beta + \sigma^\alpha\tau^\beta)(\bar\xi_\alpha\bar\eta_\beta + \bar\sigma_\alpha\bar\tau_\beta)$$
$$= (\xi^\alpha\bar\xi_\alpha)(\eta^\beta\bar\eta_\beta) + (\sigma_\alpha\bar\xi_\alpha)(\tau^\beta\bar\eta_\beta) + (\xi^\alpha\bar\sigma_\alpha)(\eta^\beta\bar\tau_\beta) + (\sigma^\alpha\bar\sigma_\alpha)(\tau^\beta\bar\tau_\beta) \qquad (75)$$

This norm clearly depends only on the equivalence class of the formal sum, and so defines a norm, and hence an inner product on $F^{\alpha\beta}$. This inner product on $F^{\alpha\beta}$ certainly satisfies **H8** and **H9** — but does it satisfy also **H10**? To show that it does involves a bit more work. A formal sum will be said to be in *normal form* if any two elements of H^α appearing in that sum are either proportional to each other (parallel) or have vanishing inner product with each other (perpendicular) and if, furthermore, any two elements of H^β appearing in that sum are also either parallel or perpendicular, and if, finally, no two terms in that sum have the property that both their H^α elements and their H^α elements are parallel. (This last condition can always be achieved by combining terms.) The norm of a formal sum in normal form is clearly positive. For example, if Eqn. (71) were in normal form, its norm would be

$$\left\|\xi^\alpha\eta^\beta + \ldots + \sigma^\alpha\tau^\beta\right\|^2 = (\xi^\alpha\bar\xi_\alpha)(\eta^\beta\bar\eta_\beta) + \ldots + (\sigma^\alpha\bar\sigma_\alpha)(\tau^\beta\bar\tau_\beta) \qquad (76)$$

Thus, the proof that the norm we have defined on $F^{\alpha\beta}$ is positive-definite will be complete if we can show that every formal sum is equivalent to a formal sum in normal form. The essential step in this demonstration is the Gram-Schmidt orthogonalization procedure. Let $\xi, \eta_1, \ldots, \eta_n$ be elements of a Hilbert space H. Then

$$\xi = \mu_1\eta_1 + \cdots + \mu_n\eta_n + \tau \qquad (\mu_i \in \mathbb{C}, (\eta_i, \tau) = 0) \tag{77}$$

That is to say, any vector in H can be written as a linear combination of $\eta_1, \ldots, \eta_n$, plus a vector perpendicular to the η's. Consider now a formal sum, say

$$\xi^\alpha\eta^\beta + \sigma^\alpha\tau^\beta + \phi^\alpha\psi^\beta \tag{78}$$

We obtain an equivalent formal sum by replacing σ^α by a vector parallel to ξ^α plus a vector perpendicular to ξ^α and combining terms. Thus, Eqn. (78) is equivalent to a formal sum

$$\xi^\alpha\eta'^\beta + \sigma'^\alpha\tau^\beta + \phi^\alpha\psi^\beta \tag{79}$$

in which σ'^α is perpendicular to ξ^α. We next obtain an equivalent formal sum by replacing ϕ^α by a vector parallel to σ'^α plus a vector parallel to ξ^α plus a vector perpendicular to both σ'^α and ξ^α. Thus, Eqn. (79) is equivalent to a formal sum

$$\xi^\alpha\eta''^\beta + \sigma'^\alpha\tau'^\beta + \phi'^\alpha\psi^\beta \tag{80}$$

with ξ^α, σ'^α, and ϕ'^α all perpendicular to each other. We now repeat this procedure with the H^β vectors to obtain a formal sum in normal form which is equivalent to Eqn. (78). Hence, every formal sum is equivalent to a formal sum in normal form, whence the norm on $F^{\alpha\beta}$ is positive-definite, whence $F^{\alpha\beta}$ is an inner-product space.

We now define the *tensor product* of H^α and H^β to be the completion of $F^{\alpha\beta}$:

$$H^\alpha \otimes H^\beta = \overline{F^{\alpha\beta}} \tag{81}$$

This is really quite complicated. An element of the tensor product is an equivalence class of Cauchy sequences in an inner-product space whose elements are equivalence classes of formal sums. Note that $F^{\alpha\beta}$ itself can be considered as a subspace of $H^\alpha \otimes H^\beta$. Elements of the tensor product which belong to $F^{\alpha\beta}$ will be called *finite* elements. In fact, we shall go one step further and consider formal sums (71) to be elements of the tensor product. Equivalent formal sums are then equal as elements of the tensor product. With these conventions, we shall be able to avoid, for the most part, having to speak always in terms of equivalence classes and Cauchy sequences.

The tensor product of more than two Hilbert spaces, $H^\alpha \otimes H^\beta \otimes \cdots \otimes H^\gamma$, is defined in a completely analogous way. We use Greek indices to indicate membership in the various tensor products, e.g., we write $\xi^{\alpha\beta\cdots\gamma}$ for a typical element of $H^\alpha \otimes H^\beta \otimes \cdots \otimes H^\gamma$. Addition and scalar multiplication within the tensor products is indicated in the obvious way:

$$\begin{aligned} \xi^{\alpha\beta\cdots\gamma} + \eta^{\alpha\beta\cdots\gamma} \\ \mu\xi^{\alpha\beta\cdots\gamma} \end{aligned} \tag{82}$$

Note, incidentally, that our original formal sums are now considerably less formal. For example, Eqn. (71) can be considered as the sum, in $H^\alpha \otimes H^\beta$, of the following elements of $H^\alpha \otimes H^\beta$: $\xi^\alpha \eta^\beta, \ldots, \sigma^\alpha \tau^\beta$.

We next observe that there is a natural, one-to-one correspondence between the formal sums which are used to obtain $H^\alpha \otimes \cdots \otimes H^\beta$,

$$\xi^\alpha \cdots \eta^\beta + \cdots + \sigma^\alpha \cdots \tau^\beta \tag{83}$$

and those which are used to obtain $\bar{H}_\alpha \otimes \cdots \otimes \bar{H}_\beta$:

$$\bar{\xi}_\alpha \cdots \bar{\eta}_\beta + \cdots + \bar{\sigma}_\alpha \cdots \bar{\tau}_\beta \tag{84}$$

That is to say, the inner product space consisting of finite elements of $\bar{H}_\alpha \otimes \cdots \otimes \bar{H}_\beta$ is the complex conjugate space of the inner product space consisting of finite elements of $H^\alpha \otimes \cdots \otimes H^\beta$. This relationship clearly continues to hold in the actual tensor product spaces. We conclude that $\bar{H}_\alpha \otimes \cdots \otimes \bar{H}_\beta$ is the same as (words we shall frequently use instead of "is naturally isomorphic with") $\overline{H^\alpha \otimes \cdots \otimes H^\beta}$. The tensor product of the complex-conjugate spaces is the same as the complex-conjugate space of the tensor product. This fact allows us to extend our index notation still further. A typical element of $\bar{H}_\alpha \otimes \cdots \otimes \bar{H}_\beta$ will be written $\bar{\xi}_{\alpha \cdots \beta}$, this being the element which corresponds to the element $\xi^{\alpha \cdots \beta}$ of $H^\alpha \otimes \cdots \otimes H^\beta$. The inner product, in the tensor product space, of two elements, $\xi^{\alpha \cdots \beta}$ and $\eta^{\alpha \cdots \beta}$, of $H^\alpha \otimes \cdots \otimes H^\beta$ can now be written as follows:

$$\xi^{\alpha \cdots \beta} \bar{\eta}_{\alpha \cdots \beta} \tag{85}$$

Thus, the index notation extends very nicely from the original collection of Hilbert spaces to the various tensor products which can be constructed.

We now introduce some operations between elements of our tensor product spaces. Let $\xi^{\alpha \cdots \beta \gamma \cdots \delta}$ and $\bar{\eta}_{\gamma \cdots \delta}$ be finite elements of $H^\alpha \otimes \ldots \otimes H^\beta \otimes H^\gamma \otimes \ldots \otimes H^\gamma$ and $\bar{H}_\gamma \otimes \ldots \otimes \bar{H}_\delta$, respectively. We can certainly associate with these two an element, $\xi^{\alpha \cdots \beta \gamma \cdots \delta} \bar{\eta}_{\gamma \cdots \delta}$ of $H^\alpha \otimes \ldots \otimes H^\beta$. For example,

$$(\xi^\alpha \eta^\beta \sigma^\gamma \tau^\delta + \phi^\alpha \psi^\beta \lambda^\gamma \kappa^\delta)(\bar{\rho}_\gamma \bar{\epsilon}_\delta + \bar{\nu}_\gamma \bar{\zeta}_\delta)$$
$$= \xi^\alpha \eta^\beta (\sigma^\gamma \bar{\rho}_\gamma)(\tau^\delta \bar{\epsilon}_\delta) + \phi^\alpha \psi^\beta (\lambda^\gamma \bar{\rho}_\gamma)(\kappa^\delta \bar{\epsilon}_\delta) + \xi^\alpha \eta^\beta (\sigma^\gamma \bar{\nu}_\gamma)(\tau^\delta \bar{\zeta}_\delta) \tag{86}$$
$$+ \phi^\alpha \psi^\beta (\lambda^\gamma \bar{\nu}_\gamma)(\kappa^\delta \bar{\zeta}_\delta)$$

Note, furthermore, that

$$\|\xi^{\alpha \cdots \beta \gamma \cdots \delta} \bar{\eta}_{\gamma \cdots \delta}\| \leq \|\xi^{\alpha \cdots \beta \gamma \cdots \delta}\| \|\bar{\eta}_{\gamma \cdots \delta}\|, \tag{87}$$

a result which is easily checked by placing $\xi^{\alpha \cdots \beta \gamma \cdots \delta}$ and $\bar{\eta}_{\gamma \cdots \delta}$ in normal form. Can this operation be extended from finite $\xi^{\alpha \cdots \beta \gamma \cdots \delta}, \bar{\eta}_{\gamma \cdots \delta}$ to the entire tensor product? To see that it can, let $\xi_i^{\alpha \cdots \beta \gamma \cdots \delta}$ and $\bar{\eta}_{i \gamma \cdots \delta}$, be Cauchy sequences of finite elements of $H^\alpha \otimes \ldots \otimes H^\beta \otimes H^\gamma \otimes \ldots \otimes H^\gamma$ and $\bar{H}_\gamma \otimes \ldots \otimes \bar{H}_\delta$, respectively. Then

$$\|\xi_i^{\alpha \cdots \delta} \bar{\eta}_{i \gamma \cdots \delta} - \xi_j^{\alpha \cdots \delta} \bar{\eta}_{j \gamma \cdots \delta}\|$$
$$= \|\xi_i^{\alpha \cdots \delta} \bar{\eta}_{i \gamma \cdots \delta} - \xi_j^{\alpha \cdots \delta} \bar{\eta}_{i \gamma \cdots \delta} + \xi_j^{\alpha \cdots \delta} \bar{\eta}_{i \gamma \cdots \delta} - \xi_j^{\alpha \cdots \delta} \bar{\eta}_{j \gamma \cdots \delta}\| \tag{88}$$
$$\leq \|\xi_i^{\alpha \cdots \delta} - \xi_j^{\alpha \cdots \delta}\| \|\bar{\eta}_{i \gamma \cdots \delta}\| + \|\xi_j^{\alpha \cdots \delta}\| \|\bar{\eta}_{i \gamma \cdots \delta} - \bar{\eta}_{j \gamma \cdots \delta}\|$$

where we have used (87). Thus, $\xi^{\alpha\cdots\beta\gamma\cdots\delta}\bar{\eta}_{\gamma\cdots\delta}$ is a Cauchy sequence in the Hilbert space $H^\alpha \otimes \ldots \otimes H^\beta$; hence it converges to some element of this Hilbert space. In this way, we associate with any (not necessarily finite) elements $\xi^{\alpha\cdots\beta\gamma\cdots\delta}$ and $\bar{\eta}_{\gamma\cdots\delta}$ an element $\xi^{\alpha\cdots\beta\gamma\cdots\delta}\bar{\eta}_{\gamma\cdots\delta}$. This product is clearly linear in the factors, and satisfies (87). The operation of contraction is thus extended from the Hilbert spaces to their tensor products. Note that (85) is now a special case.

Now let $\xi^{\alpha\cdots\beta}$ and $\eta^{\gamma\cdots\delta}$ be finite elements of $H^\alpha \otimes \ldots \otimes H^\beta$ and $H^\gamma \otimes \ldots \otimes H^\delta$, respectively. We can certainly associate with these two an element, $\xi^{\alpha\cdots\beta}\eta^{\gamma\cdots\delta}$, $H^\alpha \otimes \ldots \otimes H^\beta \otimes H^\gamma \otimes \ldots \otimes H^\delta$. For example,

$$
\begin{aligned}
(\xi^\alpha\eta^\beta + \sigma^\alpha\tau^\beta)(\phi^\gamma\psi^\delta + \lambda^\gamma\kappa^\delta) = \\
= \xi^\alpha\eta^\beta\phi^\gamma\psi^\delta + \sigma^\alpha\tau^\beta\phi^\gamma\psi^\delta + \xi^\alpha\eta^\beta\lambda^\gamma\kappa^\delta + \sigma^\alpha\tau^\beta\lambda^\gamma\kappa^\delta
\end{aligned}
\tag{89}
$$

Note, furthermore, that

$$
\|\xi^{\alpha\cdots\beta}\eta^{\gamma\cdots\delta}\| = \|\xi^{\alpha\cdots\beta}\|\|\eta^{\gamma\cdots\delta}\|
\tag{90}
$$

This operation, too, can be extended from finite elements to the entire tensor product. Let $\xi_i^{\alpha\cdots\beta}$ and $\eta_i^{\gamma\cdots\delta}$ be Cauchy sequences. Then $\xi_i^{\alpha\cdots\beta}\eta_i^{\gamma\cdots\delta}$ is also a Cauchy sequence, for

$$
\begin{aligned}
\|\xi_i^{\alpha\cdots\beta}&\eta_i^{\gamma\cdots\delta} - \xi_j^{\alpha\cdots\beta}\eta_j^{\gamma\cdots\delta}\| \\
&= \|\xi_i^{\alpha\cdots\beta}\eta_i^{\gamma\cdots\delta} - \xi_j^{\alpha\cdots\beta}\eta_i^{\gamma\cdots\delta} + \xi_j^{\alpha\cdots\beta}\eta_i^{\gamma\cdots\delta} - \xi_j^{\alpha\cdots\beta}\eta_j^{\gamma\cdots\delta}\| \\
&\leq \|\xi_i^{\alpha\cdots\beta} - \xi_j^{\alpha\cdots\beta}\|\|\eta_i^{\gamma\cdots\delta}\| + \|\xi_j^{\alpha\cdots\beta}\|\|\eta_i^{\gamma\cdots\delta} - \eta_j^{\gamma\cdots\delta}\|
\end{aligned}
\tag{91}
$$

This Cauchy sequence must converge to some element of $H^\alpha \otimes \ldots \otimes H^\beta \otimes H^\gamma \otimes \ldots \otimes H^\delta$. Thus, with any (not necessarily finite) elements $\xi^{\alpha\cdots\beta}$ and $\eta^{\gamma\cdots\delta}$ we may associate an element $\xi^{\alpha\cdots\beta}\eta^{\gamma\cdots\delta}$. This product is linear in the factors and satisfies (90). The operation of "outer product" is thus defined on our tensor product spaces. Note that the "formal products" which appear in (71) now have operational significance. (This informalization is typical of the final status of formal operations.)

We remark that our discussion of outer products above is merely the "finite part" of a more general result:

$$
(H^\alpha \otimes \ldots \otimes H^\beta) \otimes (H^\gamma \otimes \ldots \otimes H^\delta) = H^\alpha \otimes \ldots \otimes H^\beta \otimes H^\gamma \otimes \ldots \otimes H^\delta
\tag{92}
$$

That is, we have shown that finite elements of the left side can be considered as elements of the right side of (92). The rest of the proof is analogous to the proofs above.

We next consider the extension of operators from our Hilbert spaces to their tensor products. Let A be a linear operator from H^α to H^γ. This operator can be written $A^\gamma{}_\alpha$: the result of acting on $\xi^\alpha \in H^\alpha$ with A is written $A^\gamma{}_\alpha\xi^\alpha$. (The notation may be misleading here. While every element of $H^\gamma \otimes \bar{H}_\alpha$ defines such an operator, not every operator can be considered as belonging to $H^\gamma \otimes \bar{H}_\alpha$.)

With each finite element $\xi^{\alpha\cdots\beta}$ of $H^{\alpha}\otimes\ldots\otimes H^{\beta}$ we may certainly associate an element, $A^{\gamma}{}_{\alpha}\xi^{\alpha\cdots\beta}$, of $H^{\gamma}\otimes\ldots\otimes H^{\beta}$. For example,

$$A^{\gamma}{}_{\alpha}(\phi^{\alpha}\psi^{\beta} + \lambda^{\alpha}\kappa^{\beta}) = (A^{\gamma}{}_{\alpha}\phi^{\alpha})\psi^{\beta} + (A^{\gamma}{}_{\alpha}\lambda^{\alpha})\kappa^{\beta} \tag{93}$$

Unfortunately, there is in general no inequality which will permit us to extend this operation to the entire Hilbert space $H^{\alpha}\otimes\ldots\otimes H^{\beta}$. Thus, in general, $A^{\gamma}{}_{\alpha}\xi^{\alpha\cdots\beta}$ will only be defined for finite $\xi^{\alpha\cdots\beta}$; these elements, fortunately, are dense in $H^{\alpha}\otimes\ldots\otimes H^{\beta}$. There is, however, one condition under which the operation can be extended to the entire Hilbert space $H^{\alpha}\otimes\ldots\otimes H^{\beta}$. If $A^{\gamma}{}_{\alpha}$ is bounded, then

$$\|A^{\gamma}{}_{\alpha}\xi^{\alpha\cdots\beta}\| \leq |A|\|\xi^{\alpha\cdots\beta}\|. \tag{94}$$

(Proof: Use normal form.) In this case, (94) implies that $A^{\gamma}{}_{\alpha}\xi_i^{\alpha\cdots\beta}$ is a Cauchy sequence if $\xi_i^{\alpha\cdots\beta}$ is, and so we may define $A^{\gamma}{}_{\alpha}\xi^{\alpha\cdots\beta}$ for any $\xi^{\alpha\cdots\beta}$.

In fact, in many applications of the tensor product, the H^{α}, H^{β}, etc. are all merely copies of one fixed Hilbert space H. We then have one-to-one correspondences between H^{α}, H^{β}, etc. These correspondences can be indicated by retaining the root letter. For example, the element of H^{β} corresponding with $\xi^{\alpha}\in H^{\alpha}$ would be written ξ^{β}. The correspondences between our underlying Hilbert spaces induce correspondences, in an obvious way, between various tensor products. We may thus give meaning to such expressions as $\xi^{\alpha\beta} + \xi^{\beta\alpha}$. In this case — when all our underlying Hilbert spaces are copies of a Hilbert space H — we may introduce symmetrization over tensor indices (round brackets), e.g.,

$$\xi^{(\alpha\beta\gamma)} = \frac{1}{6}\left(\xi^{\alpha\beta\gamma} + \xi^{\beta\gamma\alpha} + \xi^{\gamma\alpha\beta} + \xi^{\beta\alpha\gamma} + \xi^{\gamma\beta\alpha} + \xi^{\alpha\gamma\beta}\right) \tag{95}$$

and anti-symmetrization over tensor indices (square brackets), e.g.,

$$\xi^{[\alpha\beta\gamma]} = \frac{1}{6}\left(\xi^{\alpha\beta\gamma} + \xi^{\beta\gamma\alpha} + \xi^{\gamma\alpha\beta} - \xi^{\beta\alpha\gamma} - \xi^{\gamma\beta\alpha} - \xi^{\alpha\gamma\beta}\right) \tag{96}$$

Note that any Cauchy sequence of symmetric (resp., skew) tensors of a given rank converges to a tensor which is necessarily symmetric (resp., skew). Hence, the symmetric (resp., skew) tensors of a given rank themselves form a Hilbert space. Similar remarks apply, of course, to any other symmetry on tensor indices.

There are an enormous number of facts about tensor products of Hilbert spaces. We have stated a few of them — and proven still fewer — here. It is the sheer bulk of the information, however, which makes the index notation valuable. Elementary facts are made to look elementary, and the mind is freed for important questions.

10. Fock Space: The Symmetric Case

The arena in which one discusses systems of many noninteracting identical particles is a Hilbert space called Fock space. This Hilbert space is constructed in terms of the Hilbert space H of one-particle states. Although the construction of H itself depends on the type of particle being considered (neutrinos, electrons, mesons, photons, etc.), the steps leading from H to its Fock space are independent of such details. In fact, there are two Fock spaces which can be associated with a given Hilbert space H — what we shall call the symmetric Fock space and the anti-symmetric Fock space. If H represents the one-particle states of a Boson field, the appropriate space of many-particle states is the symmetric Fock space based on H. Similarly, fermions are described by the anti-symmetric Fock space. We shall define the Fock spaces associated with a Hilbert space H and a few of the operators on these spaces.

Let H be a Hilbert space. The *(symmetric) Fock space* based on H is the Hilbert space

$$\mathbb{C} \oplus H^\alpha \oplus (H^{(\alpha} \otimes H^{\beta)}) \oplus (H^{(\alpha} \otimes H^\beta \otimes H^{\gamma)}) \oplus \cdots \qquad (97)$$

where H^α, H^β, etc. are all copies of H (Sect. 9), and where the round brackets surrounding the indices of the tensor products mean that the Hilbert space of symmetric tensors is to be used. More explicitly, an element of the symmetric Fock space consists of a string

$$\Psi = (\xi, \xi^\alpha, \xi^{\alpha\beta}, \xi^{\alpha\beta\gamma}, \ldots) \qquad (98)$$

where ξ is a complex number, ξ^α is an element of H, $\xi^{\alpha\beta}$ is a symmetric ($\xi^{\alpha\beta} = \xi^{(\alpha\beta)}$) second-rank tensor over H, $\xi^{\alpha\beta\gamma}$ is a symmetric third-rank tensor over H, etc., for which the sum

$$\|\Psi\|^2 = \xi\bar{\xi} + \xi^\alpha\bar{\xi}_\alpha + \xi^{\alpha\beta}\bar{\xi}_{\alpha\beta} + \xi^{\alpha\beta\gamma}\bar{\xi}_{\alpha\beta\gamma} + \cdots, \qquad (99)$$

which defines the norm of Ψ, converges. Physically, $\xi^{\alpha_1\cdots\alpha_n}$ represents the "n-particle contribution" to Ψ. That the tensors are required to be symmetric is a reflection of the idea that "Ψ is invariant under interchange of identical particles".

35

We next introduce the creation and annihilation operators. Let $\sigma \in H$. We associate with this σ an operator $C(\sigma)$ on Fock space, this operator defined by its action on a typical element (98):

$$C(\sigma)\Psi = (0, \sigma^\alpha \xi, \sqrt{2}\sigma^{(\alpha}\xi^{\beta)}, \sqrt{3}\sigma^{(\alpha}\xi^{\beta\gamma)}, \ldots) \tag{100}$$

Similarly, with each $\bar\tau \in \bar H$ we associate an operator $A(\bar\tau)$, defined by

$$A(\bar\tau)\Psi = (\xi^\mu \bar\tau_\mu, \sqrt{2}\xi^{\mu\alpha}\bar\tau_\mu, \sqrt{3}\xi^{\mu\alpha\beta}\bar\tau_\mu, \ldots) \tag{101}$$

This $C(\sigma)$ is called the *creation operator* (associated with σ); $A(\bar\tau)$ the *annihilation operator* (associated with $\bar\tau$). Note that the creation and annihilation operators are only defined on a dense subset of Fock space, for, in general, the sum on the right in (99) will not converge for the right sides of (100) and (101). It is an easy exercise in tensor calculus to work out the commutators of these operators:

$$\begin{aligned} [C(\sigma), C(\sigma')] &= 0 \\ [A(\bar\tau), A(\bar\tau')] &= 0 \\ [A(\bar\tau), C(\sigma)] &= (\sigma^\mu \bar\tau_\mu)\mathbb{I} \end{aligned} \tag{102}$$

For example, the last equation in (102) would be derived as follows:

$$\begin{aligned} A(\bar\tau)C(\sigma)\Psi &= A(\bar\tau)(0, \sigma^\alpha \xi, \sqrt{2}\sigma^{(\alpha}\xi^{\beta)}, \sqrt{3}\sigma^{(\alpha}\xi^{\beta\gamma)}, \ldots) \\ &= (\xi\sigma^\mu \bar\tau_\mu, \xi^\alpha \sigma^\mu \bar\tau_\mu + \sigma^\alpha \xi^\mu \bar\tau_\mu, \xi^{\alpha\beta}\sigma^\mu \bar\tau_\mu + 2\sigma^{(\alpha}\xi^{\beta)\mu}\bar\tau_\mu, \ldots) \\ C(\sigma)A(\bar\tau)\Psi &= C(\sigma)(\xi^\mu \bar\tau_\mu, \sqrt{2}\xi^{\mu\alpha}\bar\tau_\mu, \sqrt{3}\xi^{\mu\alpha\beta}\bar\tau_\mu, \ldots) \\ &= (0, \sigma^\alpha \xi^\mu \bar\tau_\mu, 2\sigma^{(\alpha}\xi^{\beta)\mu}\bar\tau_\mu, 3\sigma^{(\alpha}\xi^{\beta\gamma)\mu}\bar\tau_\mu, \ldots) \end{aligned} \tag{103}$$

Furthermore, if ϕ represents the element $(\eta, \eta^\alpha, \eta^{\alpha\beta}, \ldots)$ of Fock space, we have

$$(C(\sigma)\Psi, \phi) = (\Psi, A(\bar\sigma)\phi) \tag{104}$$

for both sides of this equation are given by the sum

$$\xi\sigma^\mu \bar\eta_\mu + \sqrt{2}\xi^\alpha \sigma^\beta \bar\eta_{\alpha\beta} + \sqrt{3}\xi^{\alpha\beta}\sigma^\gamma \bar\eta_{\alpha\beta\gamma} + \cdots \tag{105}$$

Eqn. (104) is often summarized in words by saying that $C(\sigma)$ and $A(\bar\sigma)$ are *adjoints* of each other. (An operator is thus its own adjoint if and only if it is Hermitian. Technical distinctions are sometimes made between the adjoint and the Hermitian conjugate, and between self-adjoint and Hermitian. We shall not make these distinctions until they arise.) We can now understand why the strange factors $\sqrt{2}$, $\sqrt{3}$, etc. were inserted in the definitions (100) and (101). These are the only positive real factors for which the resulting creation and annihilation operators will satisfy (102) and (104).

We next introduce two number operators. For $\sigma \neq 0$, the *number operator in the state σ* is defined by

$$N(\sigma) = \|\sigma\|^{-2}C(\sigma)A(\bar\sigma) \tag{106}$$

whence

$$N(\sigma)\Psi = \|\sigma\|^{-2}(0, \sigma^\alpha \xi^\mu \bar{\sigma}_\mu, 2\sigma^{(\alpha}\xi^{\beta)\mu}\bar{\sigma}_\mu, \ldots) \tag{107}$$

The *total number operator*, N, is defined by

$$N\Psi = (0, \xi^\alpha, 2\xi^{\alpha\beta}, 3\xi^{\alpha\beta\gamma}, 4\xi^{\alpha\beta\gamma\delta}, \ldots) \tag{108}$$

Note that these operators, too, are only defined on a dense subset of Fock space. We can think, intuitively, of N as resulting from "summing the $N(\sigma)$'s over an orthonormal basis", using

$$``\sum_{\substack{\text{orthonormal}\\\text{basis}}} \sigma^\alpha \bar{\sigma}_\mu = \delta^\alpha{}_\mu" \tag{109}$$

where $\delta^\alpha{}_\mu$ is the "Kronecker delta". The number operators, $N(\sigma)$ and N, are Hermitian. Various commutators are as follows:

$$\begin{aligned}
[N(\sigma), C(\sigma)] &= [N, C(\sigma)] = C(\sigma)\\
[N(\sigma), A(\bar{\sigma})] &= [N, A(\bar{\sigma}) = -A(\bar{\sigma})\\
[N(\sigma), N] &= 0
\end{aligned} \tag{110}$$

(Commutators involving $N(\sigma)$ are most easily evaluated using (106) and (102), rather than (107), (100), and (101). Furthermore, one can guess the commutators involving N using the intuitive remark surrounding (109).)

Finally, we write down the eigenvectors and eigenvalues of our number operators. Clearly, from (108), the eigenvalues of N are precisely the non-negative integers: $0, 1, 2, \ldots$. The most general eigenvector with eigenvalue n is:

$$(0, 0, \ldots, 0, \xi^{\alpha_1 \cdots \alpha_n}, 0, \ldots) \tag{111}$$

The eigenvectors of $N(\sigma)$ are only slightly more difficult. First note that a tensor $\xi^{\alpha_1 \cdots \alpha_n}$ satisfies

$$\sigma^{(\alpha_1}\xi^{\alpha_2 \cdots \alpha_n)\mu}\bar{\sigma}_\mu = \nu \xi^{\alpha_1 \cdots \alpha_n} \tag{112}$$

for some complex number ν if and only if

$$\xi^{\alpha_1 \cdots \alpha_n} = \sigma^{(\alpha_1} \cdots \sigma^{\alpha_m}\kappa^{\alpha_{m+1} \cdots \alpha_n)} \tag{113}$$

for some $\kappa^{\alpha_{m+1} \cdots \alpha_n}$ satisfying $\kappa^{\alpha_{m+1} \cdots \alpha_n}\bar{\sigma}_{\alpha_n} = 0$. Proof: If $\nu = 0$, we're through. If $\nu \neq 0$, (112) implies

$$\xi^{\alpha_1 \cdots \alpha_n} = \sigma^{(\alpha_1}\lambda^{\alpha_2 \cdots \alpha_n)} \tag{114}$$

for some $\lambda^{\alpha_2 \cdots \alpha_n}$. If $\lambda^{\alpha_2 \cdots \alpha_n}\bar{\sigma}_{\alpha_n} = 0$, we're through. If not, substitute (114) into (112) to obtain

$$\xi^{\alpha_1 \cdots \alpha_n} = \sigma^{(\alpha_1}\sigma^{\alpha_2}\rho^{\alpha_3 \cdots \alpha_n)} \tag{115}$$

Continue in this way. It is now clear, from (107), that the most general simultaneous eigenvector of N and $N(\sigma)$, with eigenvalues n and m, respectively, is

$$(0, 0, \ldots, 0, \sigma^{(\alpha_1} \cdots \sigma^{\alpha_m}\kappa^{\alpha_{m+1} \cdots \alpha_n)}, 0, \ldots) \tag{116}$$

where $\kappa^{\alpha_{m+1} \cdots \alpha_n}\bar{\sigma}_{\alpha_n} = 0$.

11. Fock Space:
The Anti-Symmetric Case

The definition and properties of Fock space in the antisymmetric case are closely analogous to those in the symmetric case.

Let H be a Hilbert space. The *(anti-symmetric) Fock space* based on H is the Hilbert space

$$\mathbb{C} \oplus H^\alpha \oplus (H^{[\alpha} \otimes H^{\beta]}) \oplus (H^{[\alpha} \otimes H^\beta \otimes H^{\gamma]}) \oplus \cdots \qquad (117)$$

where H^α, H^β, etc. are all copies of H, and where the square brackets surrounding the indices of the tensor products mean that the Hilbert space of anti-symmetric tensors is to be used. That is, an element of the antisymmetric Fock space consists of a string

$$\Psi = (\xi, \xi^\alpha, \xi^{\alpha\beta}, \xi^{\alpha\beta\gamma}, \ldots) \qquad (118)$$

of anti-symmetric tensors over H for which the sum

$$\|\Psi\|^2 = \xi\bar{\xi} + \xi^\alpha \bar{\xi}_\alpha + \xi^{\alpha\beta} \bar{\xi}_{\alpha\beta} + \cdots, \qquad (119)$$

which defines the norm of Ψ, converges. That the tensors are required to he anti-symmetric is a reflection of the idea that "Ψ reverses sign under the interchange of identical particles." Physically, $\xi^{\alpha_1 \cdots \alpha_n}$ represents the "n-particle contribution" to Ψ.

We associate with each $\sigma \in H$ a *creation operator*, $C(\sigma)$, and with each $\bar{\tau} \in \bar{H}$ an *annihilation operator*, $A(\bar{\tau})$, on Fock space as follows:

$$C(\sigma)\Psi = (0, \sigma^\alpha \xi, \sqrt{2}\sigma^{[\alpha}\xi^{\beta]}, \sqrt{3}\sigma^{[\alpha}\xi^{\beta\gamma]}, \ldots) \qquad (120)$$

$$A(\bar{\tau})\Psi = (\xi^\mu \bar{\tau}_\mu, \sqrt{2}\xi^{\mu\alpha}\bar{\tau}_\mu, \sqrt{3}\xi^{\mu\alpha\beta}\bar{\tau}_\mu, \ldots) \qquad (121)$$

As in the symmetric case, these operators are only defined on a dense subset of Fock space. The commutators of these creation and annihilation operators certainly exist — but they don't reduce to anything simple. We define the *anti-commutator* of two operators:

$$\{A, B\} = AB + BA \qquad (122)$$

It is the anti-commutators of the creation and annihilation operators which are simple in the antisymmetric case:

$$\{C(\sigma), C(\sigma')\} = 0$$
$$\{A(\bar{\tau}), C(\bar{\tau}')\} = 0 \tag{123}$$
$$\{A(\bar{\tau}), C(\sigma)\} = (\sigma^\mu \bar{\tau}_\mu)\mathbb{I}$$

The creation and annihilation operators are still adjoints of each other:

$$(C(\sigma)\Psi, \phi) = (\Psi, A(\bar{\sigma})\phi) \tag{124}$$

There is one further property of the creation and annihilation operators which is special to the antisymmetric case. Setting $\sigma = \sigma'$ in (123), we obtain:

$$C(\sigma)^2 = 0 \qquad A(\bar{\tau})^2 = 0 \tag{125}$$

These equations have a simple physical interpretation. If we try to create two particles in the same state, or annihilate two particles from the same state, we obtain zero. That is to say, one "can't have more than one particle in a given state." This, of course, is the essential feature of Fermi statistics.

The *number operator in the state* σ ($\neq 0$) and *total number operator* are defined by:

$$N(\sigma)\Psi = \|\sigma\|^{-2} C(\sigma) A(\bar{\sigma})\Psi = (0, \sigma^\alpha \xi^\mu \bar{\sigma}_\mu, 2\sigma^{[\alpha} \xi^{|\mu|\beta]} \bar{\sigma}_\mu, \ldots) \tag{126}$$

$$N\Psi = (0, \xi^\alpha, 2\xi^{\alpha\beta}, 3\xi^{\alpha\beta\gamma}, \ldots) \tag{127}$$

These operators are Hermitian. We can think of N as obtained by "summing the $N(\sigma)$'s over an orthonormal basis." Some commutators are:

$$[N(\sigma), C(\sigma)] = [N, C(\sigma)] = C(\sigma)$$
$$[N(\sigma), A(\bar{\sigma})] = [N, A(\bar{\sigma})] = -A(\bar{\sigma}) \tag{128}$$
$$[N(\sigma), N] = 0$$

(It is interesting that one must use commutators, and not anticommutators, to make (128) simple.) The number operator in the state σ has one further property, this one special to the antisymmetric case. From (126), (123), and (125), we have

$$N(\sigma)^2 = N(\sigma) \tag{129}$$

Clearly, (129) is again saying that "occupation numbers in the antisymmetric case are either 0 or 1." A Hermitian operator which is equal to its square is called a *projection operator*. Eqn. (129) (or, alternatively, Eqn. (126)) implies that $N(\sigma)$ is bounded (and, in fact, $|N(\sigma)| = 1$). Hence, from Section 7, $N(\sigma)$ is defined on all of Fock space. On the other hand, N is only defined on a dense subset.

Finally, we write down the eigenvectors and eigenvalues of our number operators. Once again, the eigenvalues of N are the nonnegative integers, and the general eigenvector with eigenvalue n is:

$$(0, 0, \ldots, 0, \xi^{\alpha_1 \cdots \alpha_n}, 0, \ldots) \tag{130}$$

The eigenvalue-eigenvector structure of $N(\sigma)$, however, is quite different from that of the symmetric case. In fact, (129) implies that the only eigenvalues of $N(\sigma)$ are 0 and 1. First note that

$$\sigma^{[\alpha_1}\xi^{|\mu|\alpha_2\cdots\alpha_n]}\bar{\sigma}_\mu = \nu\xi^{\alpha_1\cdots\alpha_n} \tag{131}$$

if and only if either $\xi^{\alpha_1\cdots\alpha_n}\bar{\sigma}_{\alpha_n}$ or

$$\xi^{\alpha_1\cdots\alpha_n} = \sigma^{[\alpha_1}\kappa^{\alpha_2\cdots\alpha_n]} \tag{132}$$

for some tensor $\kappa^{\alpha_2\cdots\alpha_n}$ which is antisymmetric. (We needn't require, in addition, that $\kappa^{\alpha_2\cdots\alpha_n}\bar{\sigma}_{\alpha_n} = 0$. Any multiples of σ^α which are contained in $\kappa^{\alpha_2\cdots\alpha_n}$ will be lost in (132) because of the anti-symmetrization on the right.) Thus the most general eigenvector of $N(\sigma)$ with eigenvalue 0 is (118) with

$$\xi^\alpha\bar{\sigma}_\alpha = 0, \quad \xi^{\alpha\beta}\bar{\sigma}_\alpha, \quad \xi^{\alpha\beta\gamma}\bar{\sigma}_\alpha = 0, \quad \ldots \tag{133}$$

The most general eigenvector with eigenvalue 1 is

$$(0, \sigma^\alpha\kappa, \sigma^{[\alpha}\kappa^{\beta]}, \sigma^{[\alpha}\kappa^{\beta\gamma]}, \sigma^{[\alpha}\kappa^{\beta\gamma\delta]}, \ldots). \tag{134}$$

12. Klein-Gordon Fields as Operators

Everybody knows that one essential idea of quantum field theory is that classical fields (e.g., real or complex-valued functions of position in Minkowski space) are to become operators (operator-valued functions of position in Minkowski space) on some Hilbert space. We have now assembled enough machinery to discuss this transition from fields to operators in the Klein-Gordon case. Of course, the same program will have to be carried out later — in essentially the same way — for other fields. The resulting field operators will play an important role when we discuss interactions.

Let ϕ^+ be a positive-frequency solution of the Klein-Gordon equation. That is, ϕ^+ is a complex-valued function of position in Minkowski space. The complex-conjugate function of ϕ^+, i.e., the function defined by the property that its value at a point in Minkowski space is to be the complex-conjugate of the value of ϕ^+, will be written ϕ^-. Clearly, ϕ^- is a negative-frequency solution of the Klein-Gordon equation. Finally, we introduce the real solution

$$\phi(x) = \phi^+(x) + \phi^-(x) \tag{135}$$

of the Klein-Gordon equation. The functions ϕ^+ and $\phi-$ can certainly be recovered from ϕ: they are the positive- and negative-frequency parts, respectively, of ϕ. Alternatively, these relations can be discussed in terms of functions in momentum space. Let h^+ (resp., h^-) be the complex-valued function on the future mass shell M_μ^+, (resp., the past mass shell M_μ^+) which represents the positive-frequency (resp., negative-frequency) solution ϕ^+ (resp., ϕ^-) of the Klein-Gordon equation. Then h^+ and h^- are clearly related as follows:

$$h^-(k) = \overline{h^+(-k)} \tag{136}$$

That is to say, the value of h^- at $k_a \in M_\mu^-$ is the complex-conjugate of the value of h^+ at $(-k_a) \in M_\mu^+$. Finally, the function h in momentum space which represents the real solution ϕ is given by

$$h = h^+ + h^- \tag{137}$$

This h has the property that it is invariant under simultaneous complex-conjugation and reflection through the origin, a property equivalent to the reality of ϕ.

Our ϕ^+, ϕ^-, and ϕ each assign a number (for the first two, a complex number; for the last, a real number) to each point of Minkowski space. Roughly speaking, what we want to do is find objects $\underline{\phi}^+(x)$, $\underline{\phi}^-(x)$, and $\underline{\phi}(x)$ which assign, to each point in Minkowski space, an operator. These operators are to act on the (symmetric) Fock space based on the Hilbert space of positive-frequency solutions of the Klein-Gordon equation. The relationship between the functions $\phi^+(x)$, $\phi^-(x)$, and $\phi(x)$ is to be reflected, in an appropriate way, as a relationship between the corresponding operators. Since the functions $\phi^+(x)$ and $\phi^-(x)$ are complex-conjugates of each other, we demand that the operators $\underline{\phi}^+(x)$ and $\underline{\phi}^-(x)$ be adjoints of each other; since the function $\phi(x)$ is real and the sum of $\overline{\phi}^+(x)$ and $\phi^-(x)$, we demand that the operator $\underline{\phi}(x)$ be Hermitian and the sum of the operators $\underline{\phi}^+(x)$ and $\underline{\phi}^-(x)$. Still speaking roughly, one might imagine proceeding as follows. Choose an orthonormal basis σ_i ($i = 1, 2, \ldots$) for the Hilbert space H of positive-frequency solutions of the Klein-Gordon equation. Then for each i we have a positive-frequency solution $\phi_i^+(x)$, a negative-frequency solution $\phi_i^-(x)$ (the complex-conjugate function of $\phi_i^+(x)$), and a real solution $\phi_i(x)$ (the sum of $\phi_i^+(x)$ and $\phi_i^-(x)$). Then any triple of solutions, $\phi^+(x)$, $\phi^-(x)$, and $\phi(x)$, related as above (i.e., ϕ^+ is positive-frequency, ϕ^- is its complex-conjugate, and ϕ is their sum) could be expanded in terms of our basis:

$$\phi^+(x) = \sum_i a_i \phi_i^+(x) \tag{138}$$

$$\phi^-(x) = \sum_i \bar{a}_i \phi_i^-(x) \tag{139}$$

$$\phi(x) = \phi^+(x) + \phi^-(x) = \sum_i (a_i \phi_i^+(x) + \bar{a}_i \phi_i^-(x)) \tag{140}$$

Here, $a_1, a_2, \ldots$ are simply complex numbers. Thus, triples of solutions related as above would be characterized by sequences of complex numbers. To pass from fields to operators, we could now simply replace the coefficients in the expansions (138), (139), and (140) by the corresponding creation and annihilation operators:

$$\underline{\phi}^+(x) = \hbar \sum_i \phi_i^+(x) A(\sigma_i) \tag{141}$$

$$\underline{\phi}^-(x) = \hbar \sum_i \phi_i^-(x) C(\bar{\sigma}_i) \tag{142}$$

$$\underline{\phi}(x) = \underline{\phi}^+(x) + \underline{\phi}^-(x) = \hbar \sum_i (\phi_i^+(x) A(\sigma_i) + \phi_i^-(x) C(\bar{\sigma}_i)) \tag{143}$$

Fix the basis σ_i. Then, for each position x in Minkowski space, $\phi_i^+(x)$, $\phi_i^-(x)$, and $\phi_i(x)$ are just numbers. The right sides of (141), (142), and (143) are simply (infinite!) sums of operators on Fock space. In this way, we might expect to be able to associate operators, $\underline{\phi}^+(x)$, $\underline{\phi}^-(x)$, and $\underline{\phi}(x)$ with positions in Minkowski space. These operators would, of course, satisfy the appropriate adjoint, Hermiticity, and sum conditions. One further difference between (138), (139), (140) and (141), (142), (143) should be emphasized. Whereas (138), (139), (140) involve a *particular* solution of the Klein-Gordon equation, which is expanded in terms of a basis, no such particular solution is singled out in (141), (142), (143). The sums in (141), (142), (143) need be done just once to obtain operator-valued functions on Minkowski space.

The paragraph above merely outlines a program. In order to actually carry it out, we would, at least, have to solve three problems:

i) find a basis for our Hilbert space H whose elements can be represented as smooth (or at least continuous) functions on Minkowski space (recall that an element of H is a function on the mass shell which need only be measurable and square-integrable),

ii) prove that the sums on the right of (141), (142), (143) converge in a suitable sense, and

iii) prove that the resulting operators are independent of the choice of basis.

The problem is with ii): in no reasonable sense do such sums converge. The way out of this difficulty is through the use of "smearing functions." The *support* of a function f on Minkowski space (more generally, on any topological space) is defined as the closure of the collection of points on which f is nonzero, so the support of f is a subset of Minkowski space. A smooth, real-valued function f with compact support will be called a *test function*. Note that, if G is any continuous function on Minkowski space, then the value of

$$G(f) = \int G(x)f(x)\,dV \tag{144}$$

for every test function f determines uniquely the function G. (The integral (144) converges because f has compact support.) These remarks suggest that, instead of dealing with an operator-valued function $\phi(x)$ on Minkowski space — something we have not been able to obtain — we "smear out the x-dependence of the operators":

$$\underline{\phi}(f) = \int \underline{\phi}(x)f(x)\,dV \tag{145}$$

That is to say, we consider our operators as depending, not on points in Minkowski space, but rather on the test functions themselves. Roughly speaking, $\underline{\phi}(f)$, for every test function f, has the same information content as $\underline{\phi}(x)$ for every point x of Minkowski space, just as $G(f)$ determines $G(x)$. The advantage of this formulation is that we may be able to define $\underline{\phi}(f)$, whereas we have not been able to define $\underline{\phi}(x)$. To obtain evidence on this question as to whether $\underline{\phi}(f)$ can be defined, we naively carry out the "smearing operation", (145), for (141), (142), (143):

$$\underline{\phi}^+(f) = \hbar \sum_i \left(\int \phi_i^+(x)f(x)\,dV \right) A(\sigma_i) \tag{146}$$

$$\underline{\phi}^-(f) = \hbar \sum_i \left(\int \phi_i^-(x)f(x)\,dV \right) C(\bar{\sigma}_i) \tag{147}$$

$$\underline{\phi}(f) = \underline{\phi}^+(f) + \underline{\phi}^-(f) \tag{148}$$

Thus, the operators which result from smearing are merely the creation and annihilation operators, "weighted by the components of f with respect to the basis vectors."

The preceding two paragraphs — particularly (146), (147), (148) — are intended to motivate the simple and precise definitions which follow. Let $f(x)$ be a test function. Then the Fourier inverse of f, $f'(k)$, is a smooth function on all of momentum space (not just on the mass shell: $f(x)$ doesn't have to satisfy the Klein-Gordon equation). If, however, we restrict $f'(k)$ to the future mass shell, M_μ^+, we obtain a measurable and square-integrable function on M_μ^+, and hence an element, $\sigma(f)$, of our Hilbert space. (The proof that this $f'(k)$ is actually square-integrable over the mass shell is given in books on Fourier analysis.) That is, with each test function f there is associated an element $\sigma(f)$ of H. We now define:

$$\underline{\phi}^+(f) = \hbar A(\sigma(f)) \tag{149}$$

$$\underline{\phi}^-(f) = \hbar C(\overline{\sigma(f)}) \tag{150}$$

$$\underline{\phi}(f) = \underline{\phi}^+(f) + \underline{\phi}^-(f) = \hbar(A(\sigma(f)) + C(\overline{\sigma(f)})) \tag{151}$$

(See (146), (147), (148).) These are the smeared-out field operators. We remark that these field operators are linear mappings from the real vector space of test functions to the vector space of operators on Fock space. Note that $\underline{\phi}^+(f)$ and $\underline{\phi}^-(f)$ are adjoints of each other, and that $\underline{\phi}(f)$ is Hermitian with $\underline{\phi}(f) = \underline{\phi}^+(f) + \underline{\phi}^-(f)$. It is now clear why we were not able to define operators such as $\phi^+(x)$ earlier. We can think of $\underline{\phi}^+(x)$ as being the "limit of $\underline{\phi}^+(f)$ as f approaches a δ-function at x" (see (145).) But as f approaches a δ-function, $f'(k)$ approaches a function which is not square-integrable on the mass shell. Thus, $\underline{\phi}^+(x)$ represents "creation in an un-normalizable state."

We next wish to establish a result whose intuitive content is: "in their dependence on x, the operators $\underline{\phi}(x)$, $\underline{\phi}^-(x)$, and $\underline{\phi}(x)$ satisfy the Klein-Gordon equation, e.g.,

$$(\Box + \mu^2)\underline{\phi}^+(x) = 0." \tag{152}$$

Strictly speaking, of course, (152) has no meaning, for $\underline{\phi}^+(x)$ has no meaning. Is it possible, however, to express the essential content of (152) in terms of the well-defined operators (149)? Yes. The idea is to replace (152) with the statement that the result of smearing out the left side with an arbitrary test function f gives zero:

$$\int f(x)(\Box + \mu^2)\underline{\phi}^+(x)\,dV = 0 \tag{153}$$

Formally integrating by parts twice, and throwing away surface terms since f has compact support, we are led to replace (153) (and hence (152)) by

$$\int \underline{\phi}^+(x)(\Box + \mu^2)f(x)\,dV = 0 \tag{154}$$

These considerations suggest that we replace the meaningless equations (152) or (153) by an "equivalent" equation, (154), to which we can assign a meaning.

We are thus led to the following conjecture: If f is a test function, then

$$\underline{\phi}^+((\Box + \mu^2)f) = 0$$
$$\underline{\phi}^-((\Box + \mu^2)f) = 0 \qquad (155)$$
$$\underline{\phi}((\Box + \mu^2)f) = 0$$

(Note that if f is a test function, so is $(\Box + \mu^2)f$.) In fact, our conjecture is true, for if $f'(k)$ is the Fourier inverse of $f(x)$, then $(-k_a k^a + \mu^2)f'(k)$ is the Fourier inverse of $(\Box + \mu^2)f$. But $(-k_a k^a + \mu^2)f'(k)$ vanishes on the mass shell, so $\sigma((\Box + \mu^2)f) = 0$. Eqns. (155) now follow from (149), (150), (151). We conclude that, in an appropriate sense, our operator fields satisfy the Klein-Gordon equation.

Finally, we consider the commutators of our operators. Let f and g be test functions. Since any two creation operators commute with each other, and any two annihilation operators commute with each other (see (102)), we clearly have

$$[\underline{\phi}^+(f), \underline{\phi}^+(g)] = [\underline{\phi}^-(f), \underline{\phi}^-(g)] = 0 \qquad (156)$$

Furthermore, since the commutator of any creation operator with any annihilation operator is a multiple of the identity operator (see (102)), we have

$$[\underline{\phi}^+(f), \underline{\phi}^-(g)] = \frac{\hbar}{i} D^+(f, g)\mathbb{I} \qquad [\underline{\phi}^-(f), \underline{\phi}^+(g)] = \frac{\hbar}{i} D^-(f, g)\mathbb{I} \qquad (157)$$

where $D^+(f, g)$ and $D^-(f, g)$ are complex-valued (*not* operator-valued) functions of the test functions f and g. In fact, it follows immediately from (156), (149), (150), and (102) that

$$D^+(f, g) = i\hbar\sigma^\alpha(g)\bar{\sigma}_\alpha(f)$$
$$D^-(f, g) = -i\hbar\sigma^\alpha(f)\bar{\sigma}_\alpha(f) \qquad (158)$$

Therefore,

$$D^+(f, g) = -D^-(g, f) \qquad (159)$$
$$D^+(f, g) = \overline{D^-(f, g)} \qquad (160)$$

The commutators of the ϕ operators follow from (151) and (157),

$$[\underline{\phi}(f), \underline{\phi}(g)] = \frac{\hbar}{i}(D^+(f, g) + D^-(f, g))\mathbb{I} = \frac{\hbar}{i} D(f, g)\mathbb{I} \qquad (161)$$

where the second equality is the definition of $D(f, g)$. Eqns. (159) and (160) now imply

$$D(f, g) = -D(g, f) \qquad (162)$$
$$D(f, g) = \overline{D(f, g)} \qquad (163)$$

The D-functions have one further property, which can be called *Poincaré invariance*. Let $x \to Px$ be a Poincaré transformation on Minkowski space which does not reverse the future and past time directions. (This last stipulation is necessary because the distinction between positive frequency and negative frequency requires a particular choice of a "future" time direction on Minkowski space.) Then, defining the test functions $\tilde{f}(x) = f(Px)$, $\tilde{g}(x) = g(Px)$, we have

$$D^{\pm}(\tilde{f}, \tilde{g}) = D^{\pm}(f, g) \qquad D(\tilde{f}, \tilde{g}) = D(f, g) \qquad (164)$$

The functions $D^{\pm}(f, g)$ and $D(f, g)$ are often discussed by introducing distributions $D^{\pm}(x, y)$ and $D(x, y)$ on Minkowski space, and setting

$$
\begin{aligned}
D^{\pm}(f, g) &= \int dV_x \int dV_y \, D^{\pm}(x, y) f(x) g(y) \\
D(f, g) &= \int dV_x \int dV_y \, D(x, y) f(x) g(y)
\end{aligned}
\qquad (165)
$$

It is not surprising that there should exist such distributions. A distribution, after all, is just a continuous linear mapping from the topological vector space of test functions to the reals, and $D^{\pm}(x, y)$ and $D(f, g)$ are certainly linear in their arguments. We shall not at this time discuss the topology on the space of test functions, nor prove that $D^{\pm}$ and D are continuous. Poincare invariance implies that

$$
\begin{aligned}
D^{\pm}(x, y) &= D^{\pm}(x - y) \\
D(x, y) &= D(x - y)
\end{aligned}
\qquad (166)
$$

where we have written $x - y$ for the position vector of x relative to y. It is not difficult to evaluate the functions (166) explicitly using (158) and a table of integrals. They involve Bessel functions.

There is, however, one particularly interesting property of $D(f, g)$. Test functions f and g will be said to have *relatively spacelike supports* if, for any point x of the support of f and any point y of the support of g, $x - y$ is spacelike. The property is the following: If f and g have relatively spacelike supports, then $D(f, g) = 0$. The easiest proof is by means of the distribution $D(x, y)$. Eqn. (162) implies

$$D(x - y) = -D(y - x) \qquad (167)$$

But if $x - y$ is spacelike, there is a Poincaré transformation which does not reverse future and past and which takes x to y and y to x (i.e., $x = Py$, $y = Px$). Poincaré invariance, (164), now implies

$$D(x - y) = D(y - x) \qquad (168)$$

whence $D(x, y) = 0$ for $x - y$ spacelike. That $D(f, g) = 0$ when f and g have relatively spacelike supports now follows from (165).

13. The Hilbert Space of Solutions of Maxwell's Equations

We now wish to write down the quantum theory for a system of many free (non-interacting) photons. Our starting point is the classical field equations: Maxwell's equations. The method is entirely analogous to that of the Klein-Gordon equation: the electromagnetic field plays the role of the Klein-Gordon field ϕ, the Maxwell equations the role of the Klein-Gordon equation. There are, of course, important differences between the two cases: a tensor field rather than a scalar field, two first-order tensor equations rather than one second-order scalar equation, etc. One further difference should be emphasized. Whereas the Klein-Gordon equation is, in a sense, the Schrödinger equation for a free particle, the Maxwell equations are classical (non-quantum). The electromagnetic analogy of the classical free particle, on the other hand, would be geometrical optics. Thus, we have the following table:

Electrodynamics	Free Relativistic Particle	
Geometrical Optics	Classical Dynamics	(169)
Maxwell's Equations	Klein-Gordon Equation	
Quantum Electrodynamics	Many-Particle Theory	

Theories appearing in the same row are described, mathematically, in roughly the same terms: for the first row, curves in Minkowski space; for the second row, fields in Minkowski space; for the third row, creation and annihilation operators on Fock space.

The first step is to impose the structure of a Hilbert space on a certain collection of solutions of Maxwell's equations — just as we began the second-quantization of the Klein-Gordon equation by making a Hilbert space of solutions of that equation.

The *electromagnetic field* is a skew, second-rank tensor field F_{ab} ($= F_{[ab]}$) on Minkowski space. In the absence of sources, this field must satisfy *Maxwell's equations*:

$$\nabla_{[a}F_{bc]} = 0 \tag{170}$$

$$\nabla^a F_{ab} = 0 \tag{171}$$

Eqn. (170) implies that there exists a vector field A_a on Minkowski space for which

$$F_{ab} 2\nabla_{[a} A_{b]} \tag{172}$$

Conversely, given any vector field A_a, the F_{ab} given by (172) satisfies (170). This A_a is called a *vector potential*. Substituting (172) into (171), we obtain

$$\Box A_a - \nabla_a(\nabla_b A^b) = 0 \tag{173}$$

Thus, any vector field satisfying (173) defines, via (172), a solution of Maxwell's equations, end, conversely, every solution of Maxwell's equations can be obtained from some vector potential satisfying (173). Two vector potentials, A_a and $\tilde{A}_a$, define (via (172)) the same F_{ab} if and only if

$$\tilde{A}_a = A_a + \nabla_a \Lambda \tag{174}$$

for some scalar field Λ on Minkowski space. Changes in the vector potential of the form (174) are called *gauge transformations*. By means of a gauge transformation one can find, for any solution of Maxwell's equations, a vector potential which satisfies

$$\nabla_a A^a = 0 \tag{175}$$

Vector potentials which satisfy (175) are said to be in the *Lorentz gauge*. If a vector potential for a solution of Maxwell's equations is in the Lorentz gauge, then, from (173), it satisfies

$$\Box A_a = 0 \tag{176}$$

If two vector potentials are both in the Lorentz gauge, and differ by a gauge transformation (174), then necessarily

$$\Box \Lambda = 0 \tag{177}$$

We can summarize the situation with the following awkward remark: the vector space of solutions of Maxwell's equations is equal to the quotient space of the vector space of vector fields which satisfy (175) and (176) by the vector subspace consisting of gradients of scalar fields which satisfy (177). All fields above are, of course, real.

We now do with Maxwell's equations what was done earlier with the Klein-Gordon equation: we go to momentum space. Let $A_a(x)$ be a vector potential, in the Lorentz gauge, for a solution to Maxwell's equations. Set

$$A_a(x) = \int_M A'_a(k) e^{ik_b x^b}\, dV \tag{178}$$

In (178), k represents position in momentum space, and $A'_a(k)$ associates a complex vector in momentum space with each such k. The integral on the right in (178) associates, with each point x in Minkowski space, a vector in momentum space, and hence a vector in Minkowski space at the point x. Thus, the right side of (178) defines a vector field in Minkowski space. We now demand that $A_a(x)$ given by (178) satisfy (176) and (175):

$$\Box A_a(x) = \int_M (-k_c k^c) A'_a(k) e^{ik_b x^b} \, dV = 0 \tag{179}$$

$$\nabla_a A^a(x) = \int_M i k^a A'_a(k) e^{ik_b x^b} \, dV = 0 \tag{180}$$

Eqn. (179) states that $A'_a(k)$ vanishes unless $k^a k_a = 0$. That is to say, $A'_a(k)$ need only be specified on the null cone in momentum space, or, what is the same thing, on the mass-zero shell, M_0. Thus, we can replace (178) by

$$A_a(x) = \int_{M_0} A'_a(k) e^{ik_b x^b} \, dV_0 \tag{181}$$

Eqn. (180) states that

$$k^a A'_a(k) = 0 \tag{182}$$

for every $k \in M_0$. An $A'_a(k)$ which satisfies (182) will be said to be *transverse*. Finally, the condition that $A(x)$, given by (181), be real, is

$$A'_a(-k) = \overline{A'_a(k)} \tag{183}$$

Eqn. (183) implies, in particular, that the knowledge of $A'_a(k)$ on M_0^+ determines uniquely the values of $A'_a(k)$ on M_0^-. We thus need only concern ourselves with $A'_a(k)$ on M_0^+.

To summarize, there is a one-to-one correspondence (modulo questions of convergence of Fourier integrals) between real vector fields $A_a(x)$ on Minkowski space which satisfy (175) and (176) and transverse complex vector functions $A'_a(k)$ on M_0^+.

Unfortunately, real vector fields $A_a(x)$ on Minkowski space which satisfy (175) and (176) are not the same as solutions of Maxwell's equations: we have to deal with the problem of gauge transformations. Let Λ satisfy (177), and let $\tilde{A}_a(x)$ be given by (174). Then the corresponding Fourier inverses, $\Lambda'(k)$ and $\tilde{A}'_a(k)$, are clearly related by:

$$\tilde{A}'_a(k) = A'_a(k) + i k_a \Lambda'(k) \tag{184}$$

In other words, a gauge transformation on $A_a(x)$ which preserves the Lorentz gauge corresponds simply to adding to $A'_a(k)$ a complex multiple of k_a. Note that, since k_a is null, the gauge transformations (184) do not destroy the transversality condition, (182).

To summarize, *there is a one-to-one correspondence* (modulo convergence of Fourier integrals) *between solutions of Maxwell's equations and equivalence classes of transverse complex vector functions $A'_a(k)$ on M_0^+, where two such functions $A'_a(k)$ are regarded as equivalent if they differ by a multiple of k_a.*

The reason for expressing the content of Maxwell's equations in terms of momentum space is that certain properties of the space of solutions of Maxwell's equations become more transparent there. We first impose on the (real!)

solutions of Maxwell's equations the structure of a complex vector space. To add two solutions of Maxwell's equations, one simply adds the tensor fields on Minkowski space. Expressed in terms of momentum space, this means that one adds the corresponding $A'_a(k)$. To "multiply" a solution F_{ab} of Maxwell's equations by a complex number α, one multiplies the corresponding complex vector function $A'_a(k)$ by α in the usual way, and interprets the result, $\alpha A'_a(k)$, as a solution of Maxwell's equations (necessarily, a real solution). These operations clearly extend to operations on the equivalence classes of $A'_a(k)$, and hence are well-defined operations on solutions of Maxwell's equations. It is only when α is real that multiplying a solution F_{ab} by α, in the sense described above, is equivalent to simply multiplying the tensor field F_{ab} by α in the usual way. This cannot be the case, of course, when α is complex, for the usual product, αF_{ab}, would be a complex field on Minkowski space rather than a real one, and solutions of Maxwell's equation must he real. We can, however, give a picture for what the product of i and F_{ab} ("product" and "multiply" will always refer to that operation defined above) means. Let $A_a(x)$ be a vector potential in the Lorentz space, and let $A'_a(k)$ be as in (181). Then $iA'_a(k)$ corresponds to the vector potential

$$
\int_{M_0^+} iA'_a(k)e^{ik_b x^b}\, dV_0 - \int_{M_0^-} i\,\overline{A'_a(k)}e^{ik_b x^b}\, dV_0
$$
$$
= \int_{M_0^+} A'_a(k)e^{ik_b x^b + \frac{\pi}{2}}\, dV_0 + \int_{M_0^-} \overline{A'_a(k)}e^{ik_b x^b - \frac{\pi}{2}}\, dV_0
$$

(185)

In other words, multiplication of a solution of Maxwell's equations by i corresponds to resolving F_{ab} into complex plane-waves, and shifting the phase of the positive frequency parts by $\pi/2$ while shifting the phase of the negative-frequency parts by $-\pi/2$. (In exactly the same way, the real solutions of the Klein-Gordon equation form a complex vector space.)

We next introduce an inner product on our complex vector space. We define the norm of a transverse complex vector function $A'_a(k)$ on M_0^+ by

$$
\frac{2}{\hbar}\int_{M_0^+} (-A'_a(k)\overline{A'^a(k)})\, dV_0
$$

(186)

Since $A'_a(k)$ is transverse, and since $k^a k_a = 0$ on M_0, the real number (186) is clearly invariant under gauge transformations, (184), on $A'_a(k)$. Thus, the norm (186) is well-defined on solutions of Maxwell's equations. Furthermore, the norm (186) is non-negative and vanishes when and only when $A'_a(k) = 0$ (more properly, when and only when $A'_a(k)$ is in the zero equivalence class, i.e., when and only when $A'_a(k)$ is a multiple of k_a). To prove this, we show that the integrand is non-negative. Fix k, and let

$$
A'_a(k) = m_a + in_a
$$

(187)

where m_a and n_a are real. By transversality,

$$
m_a k^a = n_a k^a = 0
$$

(188)

The integrand of (186) is

$$(-A'_a(k)\overline{A'^a(k)}) = -m^a m_a - n^a n_a \tag{189}$$

But (188) implies that m_a and n_a are either spacelike or multiples of k_a, whence (189) is nonnegative and vanishes when and only when $m_a + i n_a$ is a multiple of k_a.

Thus, the collection of all equivalence classes of (say, continuous) transverse $A'_a(k)$ on for which (186) converges has the structure of an inner-product space. Its completion is our Hilbert space, H_M, of solutions of Maxwell's equations. Just as in the Klein-Gordon case, one can describe H_M directly in momentum space. It is the collection of all equivalence classes of measurable, square-integrable (in the sense that (186) converges), transverse $A'_a(k)$ on M_0^+.

This H_M represents the one-photon states. (Intuitively, a solution of Maxwell's equations represents a "wave function" for a single photon.) The space of many-photon states is the (symmetric, since photons are bosons) Fock space based on H_M. Thus, from our earlier discussion, we have creation, annihilation, and number operators for (free) photons. The commutation relations and other properties of these operators have already been worked out.

Finally, we introduce momentum operators on H_M. Let p^a be a constant vector field in Minkowski space. Then, with each solution $F_{ab}(x)$ of Maxwell's equations, we associate another solution: multiply the solution $-\hbar p^c \nabla_c F_{ab}$ of Maxwell's equations by the number i ("multiply", of course, in the sense of H_M). We thus define a linear operator, $P(p^a)$, on H^M. In momentum space, this operator clearly takes the form

$$P(p^b)A'_a(k) = \hbar(p^b k_b)A'_a(k) \tag{190}$$

Note that the momentum operators are only defined on a dense subset of H_M, are Hermitian, and commute with each other. Another interesting property of these operators — which also holds in the Klein-Gordon case — is that "energies are positive." Let p^a be timelike and future-directed, so $P(p^a)$ represents an energy operator. Then $p^a k_a \geq 0$ for any $k_a \in M_0^+$. Hence, from (190) and (186), the inner product of σ and $P(p^a)\sigma$ is positive for any element σ ($\neq 0$) of H_M.

Although they are not commonly discussed, one can also introduce position operators on H_M. As in the Klein-Gordon case, one projects to obtain a vector field on the mass shell. Instead of taking the directional derivative of a function on the mass shell as in the Klein-Gordon case, one takes the Lie derivative of $A'^a(k)$, considered as a contravariant vector field on M_0^+. (It's important, in order to preserve transversality, that one takes $A'^a(k)$ to be a contravariant rather than a covariant field.) Finally, one includes an appropriate "divergence-type term" in the operators in order to make them be Hermitian.

14. Maxwell Fields as Operators

We shall now introduce, on the Fock space for the Maxwell equation, operators associated with the classical fields, A_a and F_{ab}, of the Maxwell theory. The definitions are closely analogous to those of the Klein-Gordon theory.

Since the classical Klein-Gordon field ϕ is a scalar field, the test functions used to "smear out" the corresponding field operators are scalar fields. In the Maxwell case, on the other hand, the classical fields are vector or tensor fields on Minkowski space. One must therefore introduce "test functions" which themselves have vectorial or tensorial character. The *support* of a tensor field $f^{a_1 \cdots a_n}$ on Minkowski space is defined as the closure of the collection of all points of Minkowski Space at which $f^{a_1 \cdots a_n} \neq 0$. A smooth, real tensor field on Minkowski space, with compact support, will be called a *test field*. In order to facilitate calculations with such test fields, it is convenient to establish the following remark:

Lemma 1. *Let $T_{a_1 \cdots a_n}$ be a smooth, totally antisymmetric tensor field on Minkowski space. Then*

$$\int T_{a_1 \cdots a_n} \nabla^{[a_1} f^{a_2 \cdots a_n]} \, \mathrm{d}V = 0 \tag{191}$$

for every totally antisymmetric test field $f^{a_2 \cdots a_n}$ if and only if

$$\nabla^{a_1} T_{a_1 a_2 \cdots a_n} = 0 \tag{192}$$

Furthermore,

$$\int T_{a_1 \cdots a_n} \nabla_m f^{m a_1 \cdots a_n} \, \mathrm{d}V = 0 \tag{193}$$

for every totally antisymmetric test field $f^{m a_1 \cdots a_n}$ if and only if

$$\nabla_{[m} T_{a_1 \cdots a_n]} = 0 \tag{194}$$

Proof. Integrating by parts once, and discarding the surface term by compact support, we have the identity

$$\int T_{a_1 \cdots a_n} \nabla^{[a_1} f^{a_2 \cdots a_n]} \, \mathrm{d}V = - \int (\nabla^{a_1} T_{a_1 \cdots a_n}) f^{a_2 \cdots a_n} \, \mathrm{d}V \tag{195}$$

for every totally antisymmetric test field $f^{a_2\cdots a_n}$. But clearly the right side of (195) vanishes for every test field if and only if (192) holds. The second part of the Lemma is proved in the same way, using the identity

$$\int T_{a_1\cdots a_n} \nabla_m f^{m a_1 \cdots a_n} \, dV = - \int (\nabla_{[m} T_{a_1 \cdots a_n]}) f^{m a_1 \cdots a_n} \, dV \qquad (196)$$

$$\square$$

Note that Lemma 1 is easily generalized to higher order equations, to other symmetries of the tensors, etc. The essential idea is that linear differential equations on a tensor field $T_{a_1\cdots a_n}$ on Minkowski space can be expressed by the condition that the smeared-out version of this field vanish for an appropriate collection of test fields.

We begin with the field operators for the vector potential. Unfortunately, the classical vector potential, $A_a(x)$, is not determined uniquely by a solution of Maxwell's equations; there is the freedom of gauge transformations (174), where Λ is a solution of the wave equation. We would expect this gauge freedom to appear in some way in the corresponding operators. The essential observation is that, by Lemma 1, the quantity

$$\int A_a f^a \, dV \qquad (197)$$

is invariant under gauge transformations provided the test field f^a is the sum of a gradient and a vector field whose divergence vanishes. Conversely, the value of the real number (197) for every test field which is the sum of a gradient and a divergence-free field determines $A_a(x)$ uniquely up to gauge transformations. We are thus led to view the gauge freedom in the vector potential as representing a restriction on the class of test fields which are appropriate for smearing out the vector potential.

The remarks above motivate the definitions below. Let f^a be a test field, and let $f'^a(k)$ be its Fourier inverse, a vector function on momentum space. Evidently, if f^a is divergence-free then

$$f'^a(k) k_a = 0 \qquad (198)$$

while if f^a is a gradient then

$$f'^a(k) = h(k) k^a \qquad (199)$$

It is clear, therefore, that if f^a is the sum of a gradient and a divergence-free field, then $f'^a(k)$, restricted to M_0^+, is transverse. In other words, we may associate, with each test field f^a on Minkowski space which is the sum of a gradient and a divergence-free field, an element $\sigma(f^a)$ of H_M. We define the vector potential operators

$$\underline{A}(f^a) = \hbar(C(\sigma(f^a)) + A(\overline{\sigma(f^a)}) \qquad (200)$$

Note that the operator (200) is Hermitian, a result one expects because the corresponding classical field is real. The definition of the electromagnetic field operators is suggested by Lemma 1 and Eqn. (172). If f^{ab} is a skew test field, we define

$$\underline{F}(f^{ab}) = \underline{A}(2\nabla_b f^{ab}) \tag{201}$$

Thus, the electromagnetic field operators (which are also Hermitian) must be smeared out with skew, second-rank test fields. (Note that the right side of (201) is well-defined, for the argument is necessarily divergence-free.)

We next verify that our field operators satisfy the same equations as the classical fields. Using Lemma 1, Eqns. (175) and (176) are translated into

$$\underline{A}(\nabla^a f) = 0 \tag{202}$$

$$\underline{A}(\Box f^a) = 0 \tag{203}$$

where f is any test function and f^a is any test field which is the sum of a gradient and a divergence-free field. Eqn. (202) follows immediately from (199). To prove (203), note that, if $f'^a(k)$ is the Fourier inverse of f^a, then $(-k_b k^b) f'^a(k)$ is the Fourier inverse of $\Box f^a$. But $(-k_b k^b) f'^a(k)$ vanishes on M_0^+, whence (203) follows. We conclude that, in a suitable sense, our vector potential operators satisfy (175) and (176). Similarly, using Lemma 1, Maxwell's equations (170) and (171) on F_{ab} are to be translated into the following conditions on the electromagnetic field operators:

$$\underline{F}(\nabla_c f^{abc}) = 0 \tag{204}$$

$$\underline{F}(\nabla^{[a} f^{b]}) = 0 \tag{205}$$

where f^{abc} is a totally antisymmetric test field and f^a is any test field. To prove (204) and (205), we substitute the definition (201):

$$\underline{F}(\nabla_c f^{abc}) = \underline{A}(2\nabla_b \nabla_c f^{abc}) = \underline{A}(0) \tag{206}$$

$$\underline{F}(\nabla^{[a} f^{b]}) = \underline{A}(2\nabla_b \nabla^{[a} f^{b]}) = \underline{A}(\nabla^a(\nabla_b f^b) - \Box f^a) \tag{207}$$

Thus, (204) is clearly true, while (205) follows immediately from (202) and (203). We conclude that, in a suitable sense, our Maxwell field operators satisfy Maxwell's equations.

Finally, we remark briefly on the commutators of the vector potential operators. Let f^a and $\tilde{f}^a$ be test fields, each of which is the sum of a gradient and a divergence-free field. Then, from (200) and (102),

$$\begin{aligned}[\underline{A}(f^a), \underline{A}(\tilde{f}^a)] &= \hbar^2(-\sigma^\alpha(f^a)\bar{\sigma}_\alpha(\tilde{f}^a) + \sigma^\alpha(\tilde{f}^a)\bar{\sigma}_\alpha(f^a))\mathbb{I} \\ &= \frac{\hbar}{i} D(f^a, \tilde{f}^a)\mathbb{I} \end{aligned} \tag{208}$$

where the second equality defines $D(f^a, \tilde{f}^a)$. Thus, $D(f^a, \tilde{f}^a)$ is real and satisfies

$$D(f^a, \tilde{f}^a) = -D(\tilde{f}^a, f^a) \tag{209}$$

These properties imply that whenever f^a and $\tilde{f}^a$ have relatively spacelike supports, $D(f^a, \tilde{f}^a) = 0$. As in the Klein-Gordon case, $D(f^a, \tilde{f}^a)$ can be written out explicitly in terms of a distribution on Minkowski space.

15. The Poincaré Group

A smooth mapping from Minkowski space to itself which preserves the norms of vectors is called a *Poincaré transformation*. If, in addition, this mapping i) does not reverse the of past and future time directions, and ii) does not reverse spatial parities (i) and ii) together are equivalent to i) and the condition that ϵ_{abcd} be invariant), then the Poincaré transformation is called a *restricted Poincaré transformation*. The result of applying two Poincaré transformations (resp., restricted Poincaré transformations) in succession is clearly again a Poincaré (resp., restricted Poincaré) transformation. These transformations thus form a group, called the *Poincaré group* (resp. *restricted Poincaré group*), $\mathscr{P}$ (resp., $\mathscr{RP}$.) One sometimes expresses this relation between the Poincaré group and Minkowski space by saying that the Poincaré group *acts* on Minkowski space. That is, we have a mapping $\Psi : \mathscr{P} \times M \to M$ ($M =$ Minkowski space) with the following properties:

$$\Psi(P, \Psi(P', x)) = \Psi(PP', x) \tag{210}$$

$$\Psi(e, x) = x \tag{211}$$

for $P, P' \in \mathscr{P}$, $x \in M$.

In fact, the Poincaré group has more structure than merely that of a group. It is also a (10-dimensional, real, differentiable) manifold. This additional manifold structure on $\mathscr{P}$ leads naturally to the notion of an "infinitesimal Poincaré transformation".

A group G which is also a smooth manifold, and for which the group operations (composition within the group, considered as a mapping from $G \times G$ to G, and the operation of taking the inverse, considered as a mapping from G to G) are smooth mappings, is called a *Lie group*. Let G be a Lie group, and let $\mathscr{L}G$ denote the collection of all contravariant vectors at the identity element e of G. This $\mathscr{L}G$ is thus a real vector space whose dimension is the same as the dimension of the manifold G. (Vectors at the identity of G represent "elements of G which differ infinitesimally from the identity.")

So far, our $\mathscr{L}G$ involves only the manifold structure of G (and, of course, the location of the identity element.) Is there some way in which the group structure of G can also be incorporated into $\mathscr{L}G$? Let $v \in \mathscr{L}G$, so v is a contravariant vector at $e \in G$. Let $g(\epsilon)$ be a smooth curve, parameterized by the parameter ϵ, in G such that $g(O) = e$ and such that the tangent vector, with respect to ϵ, of

$g(\epsilon)$ at e is just v. ("Tangent vector with respect to ϵ" means that one takes the derivative of $g(\epsilon)$ with respect to ϵ and evaluates at $\epsilon = 0$.) Similarly, let $g'(\epsilon)$ be a curve associated with $v' \in \mathbb{L}G$. Consider now the smooth curve

$$g(\epsilon)g'(\epsilon)g^{-1}(\epsilon)g'^{-1}(\epsilon) \tag{212}$$

in G. Unfortunately, the tangent vector (with respect to ϵ) of the curve (212) vanishes at e. It turns out, however, that (212) is still a smooth curve if we take as its parameter not ϵ but rather ϵ^2. The tangent vector of (212), with respect to the parameter ϵ^2, is not in general zero at e. Furthermore, this tangent vector depends only on v and v' (and not on the particular curves $g(\epsilon)$ and $g'(\epsilon)$ which actually appear in (212)), and so we may write it as follows: $[v, v']$. Thus, with any two elements, v and v', of $\mathscr{L}G$, we associate a third element, $[v, v']$, of $\mathscr{L}G$. It is by means of this bracket operation that the group structure of G appears in $\mathscr{L}G$. It can be proven that the bracket is necessarily linear, antisymmetric, and subject to the Jacobi identity:

$$\begin{aligned} [av + bv', v''] &= a[v, v''] + b[v', v''] \\ [v, av' + bv''] &= a[v, v'] + b[v, v''] \end{aligned} \tag{213}$$

$$[v, v'] = -[v', v] \tag{214}$$

$$[v, [v', v'']] + [v', [v'', v]] + [v'', [v, v']] = 0 \tag{215}$$

$(a, b \in \mathbb{R};\ v, v', v'' \in \mathscr{L}G.)$ More generally, a *Lie algebra* is a vector space V on which there is given a mapping from $V \times V$ to V (the bracket), subject to (213), (214), and (215).

To summarize, the collection $\mathscr{L}G$ of contravariant vectors at the identity of any Lie group G has the structure of a Lie algebra.

There is a more formal way of expressing the structure of $\mathscr{L}G$ in terms of that of G. Let $v \in \mathscr{L}G$. For each $g \in G$, "left multiplication by G" defines a smooth mapping from G to G which takes e to g. This mapping therefore carries v (a vector at e) to some vector at g. Repeating, for each $g \in G$, we obtain a vector field on G. That is, with each $v \in \mathscr{L}G$ there is associated a certain vector field on the manifold G. If $v, v' \in \mathscr{L}G$, then the Lie derivative of the vector field associated with v' by the vector field associated with v, evaluated at e, is precisely the element $[v, v']$ of $\mathscr{L}G$. In this formulation, properties (213), (214), and (215) of the bracket are clearly true.

The Poincaré group $\mathscr{P}$ is a Lie group: hence we have a Lie algebra $\mathscr{L}\mathscr{P}$. (The Lie algebra of the restricted Poincaré group is the same as the Lie algebra of the Poincaré group: "infinitesimal Poincaré transformations" cannot reverse past and future or spatial parities.) Fortunately, elements of $\mathscr{L}\mathscr{P}$ can be expressed very simply as certain vector fields on Minkowski space. This is not surprising: if we think of an element of $\mathscr{L}\mathscr{P}$ as representing an "infinitesimal Poincaré transformation," then its action on Minkowski space should be expressible in terms of some vector field on Minkowski space. The vector fields in Minkowski space which represent elements of $\mathscr{L}\mathscr{P}$ are those which satisfy

$$\nabla_{(a}v_{b)} = 0 \tag{216}$$

(Eqn. (216) states that the Lie derivative of the Minkowski metric by v^a vanishes.) Choosing a particular origin O, the most general solution of (216) can be expressed in the form

$$v_a = v^O_{ab} x^b + v^O_a \tag{217}$$

where v^O_a is a constant vector field on Minkowski space, v^O_{ab} is a constant skew tensor field on Minkowski space, and x^a is the position vector of x relative to our origin O. Note that the particular constant fields v^O_{ab} and v^O_a which describe a given $v_a(x)$ will depend on the choice of origin O. Note also that the dimensions are correct: six dimensions for v^O_{ab} plus four dimensions for v^O_a make ten dimensions for $\mathscr{LP}$. The bracket operation in $\mathscr{LP}$ becomes Lie derivatives of solutions of (216). That is to say, if $v, v' \in \mathscr{LP}$ correspond to solutions v_a, v'_a, respectively, of (216), then the solution of (216) which corresponds to $[v, v']$ is just

$$\mathscr{L}_v v'^a = v^b \nabla_b v'^a - v'^b \nabla_b v^a \tag{218}$$

As a check, one can verify (213), (214), and (215) for (218).

To summarize, whereas the Lie algebra $\mathscr{LP}$ of the Poincaré group arises from very general considerations involving the structure of Lie groups, $\mathscr{LP}$ can in fact be expressed very simply in terms of certain vector fields in Minkowski space.

16. Representations of the Poincaré Group

Let P be a member of the restricted Poincaré group. Then, with each positive-frequency solution $\phi(x)$ of the Klein-Gordon equation, we may certainly associate another positive-frequency solution, $\phi(Px)$. This mapping from solutions to solutions is clearly linear, and so represent an operator, U_P, on the Hilbert space H_{KG} of positive-frequency solutions of the Klein-Gordon equation. That is, for each $P \in \mathscr{RP}$, we have an operator U_P on H_{KG}. Since the operators arise from the action of $\mathscr{RP}$ on Minkowski space, we have

$$U_P U_{P'} = U_{PP'} \tag{219}$$

$$U_e = \mathbb{I} \tag{220}$$

where e denotes the identity of $\mathscr{RP}$. A mapping from a group into a collection of operators on a Hilbert space, subject to (219) and (220), is called a *representation* of the group. (More generally, the term representation is used when the operators act on any vector space.) Thus, we have defined a representation of $\mathscr{RP}$.

The inner product we have defined on H_{KG} is clearly invariant under the action of the restricted Poincaré group. That is to say, if $P \in \mathscr{RP}$, $\sigma, \tau \in H_{KG}$, we have

$$(U_P \sigma, U_P \tau) = (\sigma, \tau) \tag{221}$$

An operator on a Hilbert space which is defined everywhere and which satisfies (221) for any two elements of that Hilbert space is said to be *unitary*. A representation of a group with the property that the operator associated with each group element is unitary is called a *unitary representation*. We thus have a unitary representation of $\mathscr{RP}$ on H_{KG}.

A similar situation obtains in the Maxwell case (and for the other relativistic field equations we shall introduce later.) We have a unitary representation of $\mathscr{RP}$ on H_M.

Associated with the restricted Poincaré group $\mathscr{RP}$ is its Lie algebra $\mathscr{LP}$. What does a unitary representation of $\mathscr{RP}$ look like in terms of $\mathscr{LP}$? Let U_P be a unitary representation of the restricted Poincaré group on a Hilbert space H. Let $v \in \mathscr{LP}$, and let $P(\epsilon)$ be a corresponding curve in $\mathscr{RP}$. Consider, for each $\sigma \in H$, the right side of

$$H_v \sigma = \frac{\hbar}{i} \lim_{\epsilon \to 0} \frac{U_{P(\epsilon)} \sigma - U_{P(0)} \sigma}{\epsilon} \tag{222}$$

("lim", of course, refers to the topology on H.) It may happen, of course, that the limit in (222) does not exist for certain σ. It is normally the case in practice, however, that the limit does exist for a dense subset of H, and, furthermore, that the limit depends only on v and not on the particular curve $P(\epsilon)$. In this case, the right side of (222) is certainty linear in σ (since the U_P are), and so defines an operator H_v on H. (The factor $\hbar/i$ in (222) is for later convenience.) Thus, we associate with each $v \in \mathscr{LP}$ an operator H_v on H. The operator H_v is linear in v, i.e.,

$$H_{av+bv'} = aH_v + bH_{v'} \tag{223}$$

How is $H_{[v,v']}$ related to H_v and $H_{v'}$? To answer this question, we consider the operators associated with the curve (212):

$$U_{P(\epsilon)}U_{P'(\epsilon)}U_{P(\epsilon)}^{-1}U_{P'(\epsilon)}^{-1} \tag{224}$$

Taking the derivative (i.e., as in (222)) of (224), and evaluating at $\epsilon = 0$, we obtain the desired relation

$$[H_v, H_{v'}] = \frac{\hbar}{i}H_{[v,v']} \tag{225}$$

where we have used (222). In other words, the bracket operation on the v's becomes commutators of the H_v's. (Note that (225) is consistent with (213), (214), and (215).) One further property of the H_v's follows from the unitary character, (221), of our representation. Taking the derivative of

$$(U_{P(\epsilon)}\sigma, U_{P(\epsilon)}\tau) = (\sigma, \tau) \tag{226}$$

with respect to ϵ and evaluating at $\epsilon = 0$, we obtain, using (222),

$$(H_v\sigma, \tau) = (\sigma, H_v\tau) \tag{227}$$

That is, each operator H_v, is Hermitian.

To summarize, a unitary representation of the restricted Poincaré group on a Hilbert space H normally leads to a linear mapping from $\mathscr{LP}$ to the collection of Hermitian operators on H. The Lie bracket operation in $\mathscr{LP}$ translates to the commutator of the corresponding operators.

The general remarks above are merely intended to provide a framework for what follows. In practice, it is not necessary to go through a limiting process to obtain the Hermitian operators associated with a representation of $\mathscr{RP}$. Let $v \in \mathscr{LP}$ be the vector field v_a on Minkowski space, so v_a satisfies (216). Then, if $\phi(x)$ is a positive-frequency solution of the Klein-Gordon equation, so is the right side of

$$H_v\phi = \frac{\hbar}{i}v^a\nabla_a\phi \tag{228}$$

We thus define an operator H_v on (a dense subset of) H_{KG}. The H_v's clearly satisfy (223) and (225). In terms of momentum space, (228) may be described as follows. Let $\phi(k)$ be the Fourier inverse of $\phi(x)$ with respect to an origin O, and let v_a be given by (217) with respect to the same origin O. It then follows immediately, taking the Fourier inverse of (228), that

$$H_v\phi(k) = \hbar(v^{0a}k_a)\phi(k) + \frac{\hbar}{i}v^{0a}{}_b k^b \partial_a \phi(k) \tag{229}$$

(Note that (229) is well-defined, for $v^{0a}{}_b k^b$ tangent to M_μ^+.) We see from (229) and (30) that each H_v is Hermitian.

The situation is completely analogous for the Maxwell Hilbert space H_M. Eqn. (228) is replaced by

$$H_v F_{ab} = \frac{\hbar}{i}\mathscr{L}_v F_{ab} = \frac{\hbar}{i}(v^c \nabla_c F_{ab} + F_{cb}\nabla_a v^c + F_{ac}\nabla_b v^c) \tag{230}$$

where the multiplication by i in (230) refers to multiplication within the Hilbert space H_M. In momentum space, our Hermitian operators take the form

$$H_v A'^a(k) = \hbar(v^{0b}k_b)A'^a(k) + \frac{\hbar}{i}\mathscr{L}_{v^{0c}{}_b k^b}A'^a(k) \tag{231}$$

To summarize, we can take $\mathscr{L}\mathscr{P}$ be simply the Lie algebra of solutions of (216), and the operators H_v, to be defined by (228) and (230) (or by (229) and (231)). Then Hermiticity, (223), and (225) follow directly.

To facilitate calculations with the H_v's, it is convenient to introduce a special notation. Let $T^{a_1\cdots a_n}$ be a tensor field on Minkowski space. Then $T^{a_1\cdots a_n}$ associates, with each point x and tensor $f_{a_1\cdots a_n}$ at x, a real number, $T^{a_1\cdots a_n}f_{a_1\cdots a_n}$. For fixed x, this mapping is linear in $f_{a_1\cdots a_n}$. Furthermore, the value of this number for every x and $f_{a_1\cdots a_n}$ determines $T^{a_1\cdots a_n}$ uniquely. (Think of $f_{a_1\cdots a_n}$ as a "test function.") An operator field is what results if we replace "real number" in the remarks above by "operator on a Hilbert space H." Thus, an *operator field*, $\underline{T}^{a_1\cdots a_n}$, associates, with each point x of Minkowski space and tensor $f_{a_1\cdots a_n}$ at x, an operator on H, written $\underline{T}^{a_1\cdots a_n}f_{a_1\cdots a_n}$, such that, for x fixed, this operator is linear in $f_{a_1\cdots a_n}$. (For example, an operator field is what $\underline{A}_a(x)$ and $\underline{F}_{ab}(x)$ would be, if they existed.) Note that a tensor field is a special case of an operator field — when all the operators are multiples of the identity operator on H.

The easiest way to discuss the H_v's is as operator fields. Let x be a point of Minkowski space, and f_a a vector at x. Then the constant vector field

$$v_a = f_a \tag{232}$$

on Minkowski space certainly satisfies (216), and so defines an operator H_v (on either H_{KG} or H_M). We have defined an operator field, which we write as $\underline{P}_a$. (These, of course, are our old momentum operators, expressed in a different way.) Let x be point of Minkowski space, and let f_{ab} be a skew tensor at x. Then the vector field

$$v_a(y) = f_{ab}x^b \tag{233}$$

on Minkowski space, where x^a denotes the position vector of y relative to x, satisfies (216), and so defines an operator H_v. We have thus defined a skew operator field, which we write $\underline{P}_{ab}$.

We introduce three operations on operator fields. The first is outer product. Let f^{abc} be a tensor at the point x of Minkowski space. Write f^{abc} in the form

$$f^{abc} = m^a m^{bc} + \cdots + n^a n^{bc} \tag{234}$$

Then, for example, the outer product of $\underline{P}_a$ and $\underline{P}_{bc}$ is the operator field $\underline{P}_a\underline{P}_{bc}$, defined by

$$\underline{P}_a\underline{P}_{bc} f^{abc}{}_a = (m^a \underline{P}_a)(m^{bc} \underline{P}_{bc}) + \cdots (n^a \underline{P}_a)(n^{bc} \underline{P}_{bc}) \tag{235}$$

where the products on the right are to be interpreted as merely products of operators. Note that (235) is independent of the particular expansion (234). The outer product of two operator fields in general depends on the order in which they are written. For example, $\underline{P}_a\underline{P}_{bc} \neq \underline{P}_{bc}\underline{P}_a$. The second operation is contraction. Let f^b be a vector at the point x of Minkowski space. Then, for example, $\underline{P}^a\underline{P}_{ab}$ is the operator defined by

$$\underline{P}^a \underline{P}_{ab} f^b = (\underline{P}^c t_c)(\underline{P}_{db} t^d f^b) - (\underline{P}^c x_c)(\underline{P}_{db} x^d f^b) - (\underline{P}^c y_c)(\underline{P}_{db} y^d f^b) \\ - (\underline{P}^c z_c)(\underline{P}_{db} z^d f^b) \tag{236}$$

where t^a, x^a, y^a, z^a are vectors at x which define an orthonormal basis

$$t^a t_a = 1 \qquad x^a x_a = y^a y_a = z^a z_a = -1 \\ t^a x_a = t^a y_a = t^a z_a = x^a y_a = x^a z_a = y^a z_a = 0 \tag{237}$$

Note that (236) is independent of the choice of basis. The final operation on operator fields is differentiation. Let r^a and f^b be vectors at the point x of Minkowski space. Let x' be the point of Minkowski space whose position vector relative to x is ϵr^a, and let f'^b be f^b translated to the point x'. Then, for example, $\nabla_a\underline{P}_b$ is the operator field defined by

$$(\nabla_a\underline{P}_b)r^a f^b = \lim_{\epsilon \to 0} \frac{\underline{P}_a f'^a - \underline{P}_a f^a}{\epsilon} \tag{238}$$

(provided this limit exists). In short, operator fields are handled exactly as tensor fields, except that one must keep track of the order in products. The terms Hermitian operator field, unitary operator field, etc. are self-explanatory.

First note that $\underline{P}_a$ and $\underline{P}_{ab}$ are Hermitian operator fields. We next consider the derivatives of our two operator fields. It is clear from (232) and (238) that $\underline{P}_a$ is constant:

$$\nabla_a\underline{P}_b = 0 \tag{239}$$

To compute the derivative $\underline{P}_a b$, we first note the following fact. If $v_a(x)$, a solution of (216), is expressed in the form (217) with respect to two different origins, O and O', then

$$v_{ab}^{O'} = v_{ab}^O \qquad v_a^{O'} = v_a^O + v_{ab}^O r^b \tag{240}$$

where r^a is the position vector of O' relative to O. It now follows from (233) and (240) that, if s^{cd} is a skew tensor at O and s'^{cd} is s^{cd} translated to O', then

$$\underline{P}_{cd}s'^{cd} - \underline{P}_{cd}s^{cd} = -s_{cd}r^d\underline{P}^c \tag{241}$$

Hence,

$$\nabla_a\underline{P}_{bc} = \eta_{a[b}\underline{P}_{c]} \tag{242}$$

Eqns. (239) and (242) imply, in particular, that the second derivative of $\underline{P}_a{}^b$ vanishes. Finally, we evaluate the commutators of our operator fields. We have already seen that the momentum operators commute:

$$[\underline{P}_a, \underline{P}_b] = 0 \tag{243}$$

The other commutators are computed using the following fact: if $v_a(x)$ and $w_a(x)$ are elements of $\mathscr{LP}$, expressed in the form (217) with respect to the same origin O, then $[v, w]$ takes the form

$$2v_{[a}^{Oc}w_{b]c}^{O}x^b + (v^{Oc}w_{ac}^{O} - w^{Oc}v_{ac}^{O}) \tag{244}$$

with respect to O. Hence, from (225), (233), and (244), we have

$$[r^{ab}\underline{P}_{ab}, s^{cd}\underline{P}_cd] = \frac{2\hbar}{i}r_a{}^c s_{bc}\underline{P}^{ab} \tag{245}$$

where r^{ab} and s^{ab} are skew tensors at x. Therefore,

$$[\underline{P}_{ab}, \underline{P}_{cd}] = \frac{\hbar}{i}\left(\eta_{b[c}\underline{P}_{d]a} - (\eta_{a[c}\underline{P}_{d]b}\right) \tag{246}$$

By an identical argument, we obtain, finally,

$$[\underline{P}_a, \underline{P}_{bc}] = -\frac{\hbar}{i}\eta_{a[b}\underline{P}_{c]} \tag{247}$$

The interaction between the restricted Poincaré group and our Hilbert spaces is expressed completely and neatly by the operator fields $\underline{P}_a$ and $\underline{P}_{ab}$. The important equations on these fields are (239), (242), (243), (246), and (247).

17. Casimir Operators: Spin and Mass

Our plan is to introduce a number of relativistic field equations, and, for each one, to make a Hilbert space of an appropriate collection of solutions, to introduce the corresponding Fock space, and to replace the classical fields by operators on Fock space. This program has now been carried out for the Klein-Gordon and Maxwell equations. With each set of equations there are associated two real numbers called the mass and the spin. We could, of course, merely state what mass and what spin are to be associated with the equations in each case. It is useful, however, to see how these quantities arise in a natural way from very general considerations involving the structure of the Poincaré group. In fact, what we need of the Poincaré group is the action of its Lie algebra, $\mathscr{LP}$, on our Hilbert spaces (Sect. 16), and certain objects, called Casimir operators, associated with $\mathscr{LP}$. More generally, there are Casimir operators associated with any Lie algebra. We begin with this more general situation.

Let $\mathscr{L}$ be a Lie algebra. Then, in particular, $\mathscr{L}$ is a vector space. It is convenient to introduce an index notation. An element of $\mathscr{L}$ will be written with a raised Greek index (not to be confused with the Greek indices used in the discussion of Fock space.) Elements of the dual space of $\mathscr{L}$ (elements of the vector space of linear maps from $\mathscr{L}$ to the reals (or the complexes, if $\mathscr{L}$ were a complex vector space)) are written with lowered Greek indices. Objects with more that one index represent tensors over $\mathscr{L}$ and its dual. Finally, the action of the dual induces the operation of contraction between one raised and one lowered Greek index: this is indicated by using a repeated index. (When one wants to do anything except the most trivial calculations with multilinear algebra, it is usually simpler in the long run to introduce an index notation.) For example, the bracket operation in $\mathscr{L}$ is a bilinear mapping from $\mathscr{L} \times \mathscr{L}$ to $\mathscr{L}$, and so can be represented by a tensor $C^\mu{}_{\alpha\beta}$:

$$[v, v']^\mu = C^\mu{}_{\alpha\beta} v^\alpha v'^\beta \tag{248}$$

(This tensor is sometimes called the *structure constant tensor*.) Eqns. (214) and (215), expressed in terms of $C^\mu{}_{\alpha\beta}$, become

$$C^\mu{}_{\alpha\beta} = C^\mu{}_{[\alpha\beta]} \tag{249}$$

$$C^\mu{}_{\nu[\alpha} C^\nu{}_{\beta\gamma]} = 0 \tag{250}$$

In other words, a Lie algebra is simply a vector space over which there is given a tensor $C^\mu{}_{\alpha\beta}$, subject to (249) and (250). (For example, the invariant metric of a Lie algebra, which we shall not use here, is $C^\mu{}_{\nu\alpha}C^\nu{}_{\mu\beta}$.)

We now introduce the set $\mathscr{A}$ of all finite strings of tensors over $\mathscr{L}$:

$$(v^\alpha, v^\alpha\beta, \ldots, v^{\alpha_1 \cdots \alpha_n}, 0, 0, \ldots) \tag{251}$$

What structure do we have on $\mathscr{A}$? We can certainly add two finite strings by adding them "component-wise" — i.e., adding the vector of the first string to the vector of the second string, the second-rank tensor of the first string to the second-rank tensor of the second string, etc. — to obtain a new element of $\mathscr{A}$. Furthermore, we can multiply a finite string by a number by multiplying each element of that string by the number. Thus, $\mathscr{A}$ has the structure of an (infinite-dimensional) vector space. We can also introduce a product operation on $\mathscr{A}$. (This, in fact, is the reason for considering $\mathscr{A}$ at all.) To take the product of two finite strings, take all possible outer products consisting of one tensor from the first string and one from the second, always placing the tensor from the first string first, and add together the resulting tensors when they have the same rank to obtain the product string. For example,

$$(v^\alpha, v^{\alpha\beta}, 0, \ldots)(w^\alpha, w^{\alpha\beta}, w^{\alpha\beta\gamma}, 0, \ldots)$$
$$= (0, v^\alpha w^\beta, v^{\alpha\beta}w^\gamma + v^\alpha w^{\beta\gamma}, v^{\alpha\beta}w^{\gamma\delta} + v^\alpha w^{\beta\gamma\delta}, v^{\alpha\beta}w^{\gamma\delta\epsilon}, 0, \ldots) \tag{252}$$

Note that the product, AB, of elements A and B of $\mathscr{A}$ is linear in A and B:

$$(aA + A')B = aAB + A'B$$
$$A(aB + B') = aAB + AB' \tag{253}$$

($a \in \mathbb{R}$, $A, A', B, B' \in \mathscr{A}$.) A vector space on which there is defined a product which satisfies (253) is called an *algebra*. So $\mathscr{A}$ is an algebra. (Note that every Lie algebra is an algebra: the product is the bracket.) Since outer products of tensors are associative, so is $\mathscr{A}$:

$$A(BC) = (AB)C \tag{254}$$

($A, B, C \in \mathscr{A}$). An algebra for which (254) holds for any three of its elements is called an *associative algebra*.

Our algebra $\mathscr{A}$ so far involves only the vector-space structure of $\mathscr{L}$. (In fact, $\mathscr{A}$ is sometimes called the tensor algebra of the vector space $\mathscr{L}$.) We now want to incorporate in some way the remaining structure of $\mathscr{L}$, i.e., the bracket. Consider the collection of all elements of $\mathscr{A}$ of the form

$$(-C^\alpha{}_{\mu\nu}v^\mu w^\nu, 2v^{[\alpha}w^{\beta]}, 0, \ldots) \tag{255}$$

for $v, w \in \mathscr{L}$. Let $\mathscr{I}$ denote the set of all elements of $\mathscr{A}$ which can be written as a sum of products of elements of $\mathscr{A}$ in such a way that at least one factor in each product is of the form (255). Clearly, we have (i) $\mathscr{I}$ is a vector subspace of $\mathscr{A}$, and (ii) the product of any element of $\mathscr{I}$ with any element of $\mathscr{A}$ is again an element of $\mathscr{I}$. (A subset of an associative algebra, satisfying (i) and (ii), is called an ideal.) We now want to take the quotient algebra, $\mathscr{A}/\mathscr{I}$, of $\mathscr{A}$ by the ideal $\mathscr{I}$. We define an equivalence relation on $\mathscr{A}$: two elements of $\mathscr{A}$ are to be regarded as equivalent if their difference is in $\mathscr{I}$. That the equivalence class of any linear combination of elements A and B of $\mathscr{A}$ depends only on the equivalence classes of A and B follows from (i). That the equivalence class of the product of any two elements A and B of $\mathscr{A}$ depends only on the equivalence class of A and B follows from (ii). Thus, the collection of equivalence classes is itself an associative algebra. It is written $\mathscr{U}\mathscr{L}$ and called the *universal enveloping algebra* of the Lie algebra. To summarize, with every algebra there is associated an associative algebra $\mathscr{U}\mathscr{L}$.

There is an important relation between $\mathscr{L}$ and $\mathscr{U}\mathscr{L}$. Let $v^\alpha \in \mathscr{L}$, and let $\psi(v)$ denote the element of $\mathscr{U}\mathscr{L}$ whose equivalence class contains the element $(v^\alpha, 0, 0, \ldots)$ of $\mathscr{A}$. We thus have a — clearly linear — mapping from to $\mathscr{L}$ to $\mathscr{U}\mathscr{L}$. Furthermore, it follows from (255) that

$$\psi([v, v']) = \psi(v)\psi(v') - \psi(v')\psi(v) \tag{256}$$

for any two elements of $\mathscr{L}$. (In fact, it was to make (256) hold that we defined $\mathscr{U}\mathscr{L}$ as we did.) In other words, the bracket operation in the Lie algebra $\mathscr{L}$ corresponds to the commutator of elements of the associative algebra $\mathscr{U}\mathscr{L}$. Note that, applying ψ to both sides of (214) and (215), and using (256) and associativity, we obtain identities.

Why this interest in the universal enveloping algebra? Let $\mathscr{L}$ be a Lie algebra, and suppose, for each element v of $\mathscr{L}$, we are given an operator H_v on some fixed Hilbert space H. Suppose, furthermore, that H_v is linear in v, and that

$$H_{[v,v']} = H_v H_{v'} - H_{v'} H_v \tag{257}$$

for any $v, v' \in \mathscr{L}$. (Compare (225). It is convenient to omit the factors $\hbar/i$ — not an essential change, for such factors can always be included in the operators — when considering purely mathematical questions.) We show that this action of $\mathscr{L}$ on H can be extended naturally to an action of $\mathscr{U}\mathscr{L}$ on H. Consider an element of the associative algebra $\mathscr{A}$, written (as every element of $\mathscr{A}$ can be written) as sums of outer products of vectors, e.g.

$$(v^\alpha, u^\alpha w^\beta + p^\alpha q^\beta, r^\alpha s^\beta t^\beta, 0, 0, \ldots) \tag{258}$$

We associate with each expression of the form (258) an operator on H, e.g.,

$$H_v + H_u H_w + H_p H_q + H_r H_s H_t \tag{259}$$

It follows from the fact that H_v is linear in v that the operator (259) depends only on the element of $\mathscr{A}$ represented by (258) (and not on the particular expansion used.) Furthermore, (255) and (257) imply that if (258) is an element of $\mathscr{I}$ ($\in \mathscr{A}$), then the operator (259) is zero. Thus, (259) depends only on the equivalence class of (258). In other words, we have, for each element Γ of $\mathscr{U}\mathscr{L}$, an operator, H_Γ, on H. The operators H_Γ are linear in H:

$$H_{a\Gamma + \Gamma'} = a H_\Gamma + H_{\Gamma'} \tag{260}$$

and, clearly, satisfy

$$H_{\psi(v)} = H_v \tag{261}$$

Furthermore, it follows immediately from (259) that

$$H_{\Gamma\Gamma'} = H_\Gamma H_{\Gamma'} \tag{262}$$

Let us summarize the situation. We have a Lie algebra $\mathscr{L}$ acting on a Hilbert space H by means of the operators H_v ($v \in \mathscr{L}$) on H. The collection of all operators (at least, the collection of all those which are defined everywhere) on a Hilbert space has the structure of an associative algebra. We thus have a mapping from a Lie algebra to an associative algebra, with these two algebraic structures related via (257). Things could be better. It would be nice if we could express the bracket operation in $\mathscr{L}$ in the form

$$[v, v'] = vv' - v'v \tag{263}$$

and have

$$H_{vv'} = H_v H'_v \tag{264}$$

Then (257) would follow already from (263) and (264). This program, unfortunately, cannot be accomplished directly, for the only "product" which is defined in $\mathscr{L}$ is the entire bracket, and not the individual terms on the right of (263). But it can be accomplished indirectly. We "enlarge" $\mathscr{L}$ to $\mathscr{UL}$. We still cannot write (263) — but instead we have (256). (Eqn. (256) also states that the algebraic structure of $\mathscr{L}$ has been incorporated into that of $\mathscr{UL}$.) We still cannot write (264) — but instead we have (262). In short, since $\mathscr{L}$ is being mapped to an associative algebra (the operators on H), and since the natural thing to map to an associative algebra is another associative algebra, we "force associativity" on $\mathscr{L}$ by enlarging it to $\mathscr{UL}$.

We can now introduce the Casimir operators. A *Casimir operator* of the Lie algebra $\mathscr{L}$ is an element of the center of $\mathscr{UL}$, i.e., an element Γ of $\mathscr{UL}$ such that

$$\Gamma\Delta - \Delta\Gamma = 0 \tag{265}$$

for every $\Delta \in \mathscr{UL}$. It should be emphasized that the Casimir operators of a Lie algebra are not themselves elements of that Lie algebra, but rather of its universal enveloping algebra. That is, they must be represented as strings of tensors over $\mathscr{L}$. Note that the collection of all Casimir operators of a Lie algebra form an associative algebra. Finally, we remark that the universal enveloping algebra $\mathscr{UL}$ and hence the Casimir operators (which are not operators, as we have defined them, but merely elements of an algebra) are fixed once and for all given the Lie algebra $\mathscr{L}$. They do not depend on the presence of a Hilbert space H or on the H_v's. For example, the Casimir operators of $\mathscr{LP}$ (the Lie algebra of the Poincaré group) simply exist. (In fact, there are just two algebraically independent ones.) They needn't be found individually for H_{KG}, H_M, etc.

Now suppose again that we have a Hilbert space H and, for each $v \in \mathscr{L}$, an operator H_v on H, where the H_v's are linear in v and satisfy (257). Then we have an operator H_Γ on H for each $\Gamma \in \mathscr{UL}$, and, in particular, an operator on H for each Casimir operator of $\mathscr{L}$. Eqns. (265) and (262) imply that, if Γ is a Casimir operator, then H_Γ commutes with all the H_Δ's, and, in particular, with all the H_v's (see (261)). This normally implies, as we shall see in examples later, that H_Γ is simply a multiple of the identity operator. Thus, the Casimir operators assign numbers to our relativistic field equations, i.e., to our representations of the restricted Poincaré group.

The words "normally implies" above are rather vague. I do not know whether or not there is a general theorem which implies that the Casimir H_Γ's are multiples of the identity in cases of interest. However, the following result suggests this conclusion:

Lemma 2 (Schur's Lemma). *Let H be a finite-dimensional complex vector space, and let $\mathscr{L}$ be a set. Suppose, for each $v \in \mathscr{L}$, we are given an operator (defined everywhere) on H, H_v. Suppose, furthermore, that the only vector subspaces S of H having the property that $H_v \sigma \in S$ for every $v \in \mathscr{L}$ and $\sigma \in S$ are $S = \{0\}$ and $S = H$. Let K be an operator (defined everywhere) on H which commutes with all the H_v's. Then K is some complex multiple of the identity.*

Proof. Since H is a complex vector space, K has at least one eigenvector, i.e., there exists a complex number κ and a nonzero element σ of H such that

$$K\sigma = \kappa\sigma \qquad (266)$$

Fix κ, and let S be the collection of all σ's which satisfy (266). Then, for $\sigma \in S$, $v \in \mathscr{L}$,

$$K(H_v \sigma) = H_v K\sigma = \kappa(H_v \sigma) \qquad (267)$$

Hence, $H_v \sigma \in S$. By hypothesis, therefore, $S = \{0\}$, or $S = H$. But by construction S contains at least one nonzero element of H, so we must have $S = H$. In other words, every element of H satisfies (266), whence $K = \kappa I$. $\square$

We now want to apply all this mathematics to our relativistic fields. As usual, one can regard the formal developments as merely providing motivation and insight into what turn out to be very simple notions in practice. The operators on our Hilbert spaces associated with the Casimir operators of $\mathscr{L}\mathscr{P}$ can be expressed quite easily in terms of the operator fields $\underline{P}_a$ and $\underline{P}_{ab}$ discussed in Section 16. The first Casimir operator is

$$\underline{P}_a \underline{P}^a = m^2 \qquad (268)$$

We see from (243) and (247) that m^2 commutes with $\underline{P}_a$ and $\underline{P}_{ab}$. Furthermore, (239)) implies that m^2 is a constant operator field. Hence, m^2 is just an ordinary operator on our Hilbert spaces. It turns out to be a multiple of the identity (as suggested above), and that multiple is called the (squared) *mass* of the field. To define the second Casimir operator, we first introduce the operator field

$$\underline{W}^a \epsilon^{abcd} \underline{P}_b \underline{P}_{cd} \qquad (269)$$

Then (239) and (242) imply that $\underline{W}^a$ is constant. The second Casimir operator is the left side of

$$\underline{W}_a \underline{W}^a = -\hbar^2 m^2 s(s+1) \qquad (270)$$

Note, from (243), (246), and (247), that $\underline{W}_a\underline{W}^a$ commutes with $\underline{P}_a$ and $\underline{P}_{ab}$. It turns out to be a multiple of the identity, and the non-negative number s which makes that multiple be the right side of (270) is called the *spin* of the field. We remark that the mass and spin are associated not with each individual solution of a relativistic field equation, but rather with the equation itself.

Unfortunately, (270) will not give the spin s when $m = 0$. In the massless case, it is found that there is a number s for which

$$\underline{W}_a = s\hbar\underline{P}_a \tag{271}$$

and so this equation is used to define the spin. This definition has an interesting consequence. Note that the definition of $\underline{W}^a$ involves one ϵ^{abcd}, while there are none in $\underline{P}_a$. That is, $\underline{W}^a$ is a pseudovector, while $\underline{P}^a$ an ordinary vector (operator field). Hence, the spin s is a pseudoscalar in the massless case, and a scalar when $m \neq 0$. We shall see shortly that this feature is related to the notion of helicity.

Finally, we evaluate the mass and spin in the Klein-Gordon and Maxwell cases. Let r^a be a vector at the point x of Minkowski space, and let positive-frequency solution of the Klein-Gordon equation. Then

$$r^a\underline{P}_a(r^b\underline{P}_b\phi) = r^a\underline{P}_a\left(\frac{\hbar}{i}r^b\nabla_b\phi\right) = -\hbar^2 r^a r^b\nabla_a\nabla_b\phi \tag{272}$$

To evaluate $\underline{P}_a\underline{P}^a\phi$ we must sum (272), with the appropriate signs, as r^a runs over an orthonormal tetrad (see (236).) Clearly, the result of taking this sum is simply to replace $r^a r^b$ by the Minkowski metric, η^{ab}. So

$$\underline{P}_a\underline{P}^a\phi = -\hbar^2\Box\phi = \hbar^2\mu^2\phi \tag{273}$$

But $\hbar^2\mu^2$ for the Klein-Gordon equation is what we earlier (c.f. (5)) called m^2. Hence, the m in Sect. 1 is indeed the mass for the Klein-Gordon equation. To evaluate the spin, let r^a be a vector and s^{ab} a skew tensor at the point x. Then, writing x^a for the position vector relative to x,

$$r^a\underline{P}_a(s^{bc}\underline{P}_{bc}\phi) = r^a\underline{P}_a\left(\frac{\hbar}{i}s^b{}_c x^c\nabla_b\phi\right) = \frac{\hbar}{i}r^a\nabla_a\left(\frac{\hbar}{i}s^b{}_c x^c\nabla_b\phi\right)$$
$$= -\hbar^2(r^c s^b{}_c\nabla_b\phi + r^a s^b{}_c x^c\nabla_a\nabla_b\phi) \tag{274}$$

Let u^a be another vector at x. Then, to evaluate $u_a\underline{W}^a\phi$, we must sum (274) over r's and s's so that $\sum r^b s^{cd} = u_a\epsilon^{abcd}$. The result, clearly, is just to replace the combination $r^b s^{cd}$ in (274) by $u_a\epsilon^{abcd}$. So,

$$u_a\underline{W}^a\phi = -\hbar^2 u_a\epsilon^{abcd}(\eta_{bd}\nabla_c\phi + x_d\nabla_b\nabla_c\phi) = 0 \tag{275}$$

Thus, $\underline{W}^a$ is zero on H_{KG}. Now (270) implies $s = 0$ in the massive case, while (271) gives $s = 0$ in the massless case. The Klein-Gordon equation describes a particle of mass m and spin zero.

It is enlightening, instead of treating just the Maxwell case, to discuss the more general equation

$$(\Box + \mu^2)A_a = 0 \qquad \nabla^a A_a = 0 \qquad (276)$$

Maxwell's equations are obtained for $\mu = 0$. If r^a is a vector at x,

$$r^a \underline{P}_a(r^b \underline{P}_b A_c) = -\hbar^2 r^a r^b \nabla_a \nabla_b A_c \qquad (277)$$

Hence,

$$\underline{P}_a \underline{P}^a A_b = -\hbar^2 \Box A_b = \hbar^2 \mu^2 A_b \qquad (278)$$

Hence, the mass of the fields described by (276) is just as in the Klein-Gordon case. (In particular, photons have mass zero.) If r^a is a vector and s^{ab} a skew tensor at x, then

$$
\begin{aligned}
r^a \underline{P}_a(s^{bc} \underline{P}_{bc} A_d) &= r^a \underline{P}_a \left(\frac{\hbar}{i} \mathcal{L}_{s^b{}_c x^c} A_d \right) \\
&= \frac{\hbar}{i} r^a \nabla_a \left(\frac{\hbar}{i} \left(s^b{}_c x^c \nabla_b A_d + A_b s^b{}_d \right) \right) \\
&= -\hbar^2 \left[r^a s^b{}_c x^c \nabla_a \nabla_b A_d + r^c s^b{}_c \nabla_b A_d + r^a s^b{}_d \nabla_a A_b \right] \quad (279)
\end{aligned}
$$

Therefore, by the same argument as before,

$$
\begin{aligned}
u_a \underline{W}^a A_e &= -\hbar^2 u_a \epsilon^{abcd} \left[x_d \nabla_b \nabla_c A_e + \eta_{bd} \nabla_c A_e + \eta_{ed} \nabla_b A_c \right] \\
&= -\hbar^2 \epsilon_{aebc} u^a \nabla^b A^c
\end{aligned}
\qquad (280)
$$

where u_a is a vector at x. Hence,

$$
\begin{aligned}
\underline{W}_a \underline{W}^a A_c &= -\hbar^2 \epsilon_{aebc} \nabla^b \left(-\hbar^2 \epsilon^{acpq} \nabla_p A_q \right) \\
&= -4\hbar^4 \nabla^b (\nabla_{[e} A_{b]}) = -2\hbar^4 \mu^2 A_e
\end{aligned}
\qquad (281)
$$

Thus, the spin of the fields (276) is $s = 1$, provided $\mu \neq 0$.

But something appears to be wrong in the Maxwell case, $\mu = 0$. Eqn. (280) is not proportional to

$$u_a \underline{P}^a A_e = \frac{\hbar}{i} r^a \nabla_a A_e \qquad (282)$$

First note that, by a gauge transformation on the right in (282), we can write

$$u_a \underline{P}^a A_e = \frac{\hbar}{2i} u^a \nabla_{[a} A_{e]} \qquad (283)$$

We still don't have proportionality with (280). The reason is that the representation of $\mathcal{L}\mathcal{P}$ on H_M is not irreducible. A solution of Maxwell's equations is said to have *positive* (resp. *negative*) *helicity* if

$$\epsilon^{abcd} F_{cd} = \pm \frac{i}{2} F^{ab} \qquad (284)$$

with the plus (resp. minus) sign on the right. (In (284), "i" means multiplication in H_M. The factor $i/2$ is necessary because $\epsilon^{abcd}\epsilon_{cdef}F^{ef} = -4F^{ab}$ for any skew F^{ab}.) In momentum space, a positive-helicity or negative-helicity solution takes the form

$$A'_a(k) = m_a + in_a \tag{285}$$

with $m^a m_a = n^a n_a$, $m^a n_a = 0$. The two helicities arise because there are two directions through which m_a can be rotated through $90°$ to obtain n_a. Every solution of Maxwell's equations can be written uniquely as the sum of a positive and a negative helicity solution. Furthermore, the inner product of a positive helicity solution with a negative helicity solution is zero. (These facts follow immediately from (285).) Thus, H_M is the direct sum of the Hilbert space of positive-helicity solutions with the Hilbert space of negative-helicity solutions. On the Hilbert space of positive-helicity solutions, $s = 1$; on the Hilbert space of negative-helicity solutions, $s = -1$.

18. Spinors

Particles with half-integer spin (electrons, neutrinos, etc.) are described by mathematical objects called spinor fields. We shall base our treatment of such particles on what are called "two-component spinors" (rather than the more common four-component spinors.) Essentially the only difference between the two is one of notation. Whereas the two-component spinors lend themselves more naturally to an index notation, the four-component spinors are slightly more convenient when discussing discrete symmetries. We shall first define (two-component) spinors, and then indicate how formulae can be translated to the four-component language.

Let $\mathfrak{C}$ be a two-dimensional, complex vector space. Membership in $\mathfrak{C}$ will be indicated with a raised, upper case Latin index, e.g., ξ^A, η^A, etc. We introduce three additional two-dimensional complex vector spaces:

i) The complex-conjugate space, $\bar{\mathfrak{C}}$, of $\mathfrak{C}$ (see Sect. 8);

ii) The dual space, $\mathfrak{C}^*$, of $\mathfrak{C}$ (i.e., the vector space of linear mappings from $\mathfrak{C}$ to the complexes; and

iii) The complex-conjugate space of the dual space of $\mathfrak{C}$, $\bar{\mathfrak{C}}^*$ (or, what is the same thing, the dual space of the complex-conjugate space of $\mathfrak{C}$).

Membership in $\bar{\mathfrak{C}}$ will be indicated with a primed raised index, e.g., $\xi^{A'}$; membership in $\mathfrak{C}^*$ by an unprimed lowered index, e.g., ξ_A; and membership in $\bar{\mathfrak{C}}^*$ by a primed lowered index, e.g., $\xi_{A'}$. That is, we now have four different vector spaces, with their elements represented by four different index combinations. What operations can be performed on the elements of $\mathfrak{C}$, $\bar{\mathfrak{C}}$, $\mathfrak{C}^*$, and $\bar{\mathfrak{C}}^*$? Of course, we can multiply elements by complex numbers, and add elements which belong to the same vector space (i.e., which have the same index structure). Furthermore, our four vector spaces can be grouped into pairs which are complex-conjugates of each other: $\mathfrak{C}$ and $\bar{\mathfrak{C}}$ are complex-conjugates of each other, and $\mathfrak{C}^*$ and $\bar{\mathfrak{C}}^*$ are complex-conjugates of each other. Thus, we have an operation of complex-conjugation, which, applied to an element of $\mathfrak{C}$ (resp. $\bar{\mathfrak{C}}$, $\mathfrak{C}^*$, or $\bar{\mathfrak{C}}^*$), yields an element of $\bar{\mathfrak{C}}$ (resp. $\mathfrak{C}$, $\bar{\mathfrak{C}}^*$, $\mathfrak{C}^*$). For example,

$$\overline{\xi^A} = \bar{\xi}^{A'} \qquad \overline{\eta^{A'}} = \bar{\eta}^A$$
$$\overline{\sigma_A} = \bar{\sigma}_{A'} \qquad \overline{\tau_{A'}} = \bar{\tau}_A \tag{286}$$

Note the effect of the operation of complex-conjugation on the index structure: it adds a prime if there was none before, and a deletes a prime if there was one before. Finally, we can group our four vector spaces into pairs which are duals of each other: $\mathfrak{C}$ and $\mathfrak{C}^*$ are duals of each other, and $\bar{\mathfrak{C}}$ and $\bar{\mathfrak{C}}^*$ are duals of each other. We thus have the operation of contraction: an element $\xi^A \in \mathfrak{C}$ together with an element $\eta_A \in \mathfrak{C}^*$ defines a complex number, $\xi^A \eta_A$; an element $\sigma^{A'} \in \bar{\mathfrak{C}}$ together with an element $\tau_{A'} \in \bar{\mathfrak{C}}^*$ defines a complex number, $\sigma^{A'} \tau_{A'}$. One indicates contraction, as above, by using a repeated index. Note that one can only contract between a raised and a lowered index when these are of the same type (both primed or both unprimed). We have, for example,

$$\overline{(\xi^A \eta_A)} = \bar{\xi}^{A'} \bar{\eta}_{A'} \tag{287}$$

(Note that the index notation we used for Hilbert spaces is essentially a special case of that described above. The inner product on a Hilbert space induces a natural isomorphism between $\bar{\mathfrak{C}}$ and $\mathfrak{C}^*$, and between $\bar{\mathfrak{C}}^*$ and $\mathfrak{C}$. One can therefore do away with primed indices entirely.)

Now consider the various tensor products between $\mathfrak{C}$, $\bar{\mathfrak{C}}$, $\mathfrak{C}^*$, and $\bar{\mathfrak{C}}^*$. The particular tensor product to which an object belongs is indicated by its index structure, e.g., $T^{A\cdots BC'\cdots D'}{}_{E\cdots FG'\cdots H'}$. Complex-conjugation extends in an obvious way to the tensor products, e.g.,

$$\overline{T^{A\cdots BC'\cdots D'}{}_{E\cdots FG'\cdots H'}} = \bar{T}^{A'\cdots B'C\cdots D}{}_{E'\cdots F'G\cdots H} \tag{288}$$

We define a *spinor-space* as a two-dimensional, complex vector space $\mathfrak{C}$ on which is specified a nonzero object ϵ_{AB} which is skew:

$$\epsilon_{AB} = \epsilon_{[AB]} \tag{289}$$

(Note that, since $\mathfrak{C}$ is two-dimensional, any two skew ϵ_{AB}'s differ at most by a complex factor. Hence, there is "just one" spinor space.) Elements of the tensor products will be called *spinors*. Thus, we can multiply spinors by complex numbers, add spinors when they have the same index structure, take outer products of spinors, and contract over spinor indices (one raised and one lowered, both primed or both unprimed.) Since $\epsilon_{AB} \neq 0$, there is a unique spinor ϵ^{AB} which is skew and satisfies

$$\epsilon_{AM} \epsilon^{BM} = \delta^B{}_A \tag{290}$$

where $\delta^B{}_A$ is the unit spinor (defined by $\delta^B{}_A \xi^A = \xi^B$ for all ξ^A.) We can now "raise and lower spinor indices" (i.e., define isomorphisms between $\mathfrak{C}$ and $\mathfrak{C}^*$ and between $\bar{\mathfrak{C}}$ and $\bar{\mathfrak{C}}^*$):

$$\xi^A = \epsilon^{AB} \xi_B \qquad\qquad \xi_A = \xi^B \epsilon_{BA}$$
$$\eta^{A'} = \bar{\epsilon}^{A'B'} \eta_{B'} \qquad\qquad \eta_{A'} = \eta^{B'} \bar{\epsilon}_{B'A'} \tag{291}$$

(Note the placement of indices in (291).) Similarly, one can raise and lower an index of a spinor with more than one index. Note that, since ϵ_{AB} is skew, we have

$$\xi^A \sigma_A = -\xi_A \sigma^A \tag{292}$$

Let V denote the collection of all spinors $\xi^{AA'}$ which are real:

$$\overline{\xi^{AA'}} = \xi^{AA'} \tag{293}$$

We can certainly add two elements of V obtain another element of V, and multiply an element of V by a real number to obtain another element of V. (Note that multiplication by a complex number does not preserve (293): $\overline{(\alpha\xi^{AA'})} = \bar{\alpha}\overline{\xi^{AA'}}$.) Thus, V is a real vector space, which, as can easily be checked by introducing a basis, is four-dimensional. (The collection of all Hermitian 2×2 matrices is a real four-dimensional vector space.) We can furthermore define on V an inner product (i.e., a metric) via

$$\xi^{AA'}\xi_{AA'} \tag{294}$$

By introducing a basis, or otherwise, it is easily checked that the signature of this metric is $(+, -, -, -)$.

So far, spinor space is a purely mathematical construct. In order to actually use the spinor space in physics, we must somehow tie it down to our space-time — Minkowski space. This is accomplished, of course, through the vector space V of solutions of (293). The vectors at a point x of Minkowski space form a four-dimensional vector space on which there is a metric of signature $(+, -, -, -)$. We "tie down" spinor space, therefore, by specifying some metric-preserving isomorphism between V and this vector space of vectors at the point x. We assume that such an isomorphism has been fixed once and for all. Thus, we can regard a tensor in Minkowski space at x, e.g., $T^{ab}{}_c$, defining a spinor, $T^{AA'BB'}{}_{CC'}$ which is real:

$$\overline{T^{AA'BB'}{}_{CC'}} = T^{AA'BB'}{}_{CC'} \tag{295}$$

We shall allow ourselves to write such "equivalent" quantities as equal:

$$T^{ab}{}_c = T^{AA'BB'}{}_{CC'} \tag{296}$$

In other words, we are free to replace a lowercase Latin index (tensor index in Minkowski space) by the corresponding uppercase Latin index written twice, once unprimed and once primed. For example, the metric in Minkowski space takes the form (see (294))

$$\eta_{ab} = \epsilon_{AB}\bar{\epsilon}_{A'B'} \tag{297}$$

We may thus regard tensors at x as merely a special case of spinors (those having an equal number of primed and unprimed indices, and which, for a real tensor, are real). Finally, note that translation in Minkowski space defines a metric-preserving isomorphism between the vectors at x and the vectors at any other point y. Hence, we have automatically spinors at the point y. More generally, we have the notion of a spinor field, a spinor function of position in Minkowski space. Tensor fields in Minkowski space are, of course, a special case. We can multiply spinor fields by real or complex scalar fields, and spinor fields (when they have the same index structure), take outer products of spinor fields, and contract appropriate spinor indices.

It is possible, in addition, to take derivatives of spinor fields. Let, for example, $T^A{}_{BC'}$ be a spinor field. Let r^m be a vector at x, and let x' be the point whose position vector relative to x is ϵr^m. We define $\nabla_m T^A{}_{BC'}$ by

$$r^m \nabla_m T^A{}_{BC'} = \lim_{\epsilon \to 0} \frac{T^A{}_{BC'}(x') - T^A{}_{BC'}(x)}{\epsilon} \tag{298}$$

The replacement of Minkowski tensor indices by spinor indices can, of course, be extended to the index of the derivative operator. That is,

$$\nabla_m T^A{}_{BC'} = \nabla_{MM'} T^A{}_{BC'} \tag{299}$$

In short, the mechanics of calculating with spinor fields is in no essential way different from that of tensor fields. The one point one has to be careful about is the index locations in contractions (see (292).)

One further question must be discussed. To what extent does the imposition of the notion of spinor fields on Minkowski space enrich the structure of Minkowski space? That is, are there essentially inequivalent spinor structures on Minkowski space? To obtain evidence on this question, consider the collection of vector fields on Minkowski space of the form

$$\xi^A \bar{\xi}^{A'} \tag{300}$$

where $\xi^A(x)$ is a spinor field. This vector field is certainly real, and, from (294) is null. (From (292), $\xi^A \xi_A = 0$.) Furthermore, the inner product of two such fields,

$$(\xi^A \bar{\xi}^{A'})(\eta_A \bar{\eta}_{A'}) = (\xi^A \eta_A)(\bar{\xi}^{A'} \bar{\eta}_{A'}) = (\xi^A \eta_A)\overline{(\xi^B \eta_B)} \tag{301}$$

is necessarily non-negative. Thus, the spinor structure on Minkowski space determines a particular time orientation, which we may specify as being the "future." (Past-directed null vectors then have the form $-\xi^A \bar{\xi}^{A'}$.) Furthermore, the tensor field on Minkowski space defined by the right side of

$$\epsilon_{abcd} = -i(\epsilon_{AB}\epsilon_{CD}\bar{\epsilon}_{A'C'}\bar{\epsilon}_{B'D'} - \bar{\epsilon}_{A'B'}\bar{\epsilon}_{C'D'}\epsilon_{AC}\epsilon_{BD}) \tag{302}$$

is real, totally antisymmetric, and satisfies $\epsilon^{abcd}\epsilon_{abcd} = -24$. Hence, this must be all alternating tensor on Minkowski space. Thus, a spinor structure on Minkowski space induces both temporal and spatial parities on Minkowski space. In fact, this is all the structure induced on Minkowski space by a spinor structure. More precisely, given two metric-preserving isomorphisms between V and vectors in Minkowski space, such that these induce the same spatial and temporal parities on Minkowski space, these are related by a linear mapping from $\mathfrak{C}$ onto $\mathfrak{C}$ which preserves ϵ_{AB} (i.e., by an element of SL(2, $\mathbb{C}$).) Finally, note that there are precisely two ϵ_{AB}-preserving linear mappings on $\mathfrak{C}$ which leave V (and hence tensors in Minkowski space) invariant, namely the identity and minus the identity. This is the statement of the "two-valuedness" associated with spinors.

Finally, we briefly indicate how one translates formulae from the two-component to the four-component spinor notation. A four-component spinor is a pair of two-component spinors, $(\xi^A, \eta_{A'})$, consisting of one spinor with a primed and one with an unprimed index. One then normally chooses a basis for $\mathfrak{C}$ and writes this pair out as a 4×1 column matrix. The "γ-matrices" in the four-component notation serve the function of combining these components in the appropriate way to obtain the various scalar, vector, and tensor fields on Minkowski space associated with the pair $(\xi^A, \eta_{A'})$. For example, a pair $(\xi^A, \eta_{A'})$ defines the following fields on Minkowski space:

$$\xi^A \bar{\eta}_A, \quad \xi_A \bar{\xi}_{A'}, \quad \eta_{A'} \bar{\eta}_A, \quad \xi_A \eta_{A'}, \quad \xi_A \xi_B \bar{\epsilon}_{A'B'},$$
$$\eta_{A'} \eta_{B'} \epsilon_{AB}, \quad \xi_A \bar{\eta}_B \bar{\epsilon}_{A'B'}, \quad \xi_B \bar{\xi}_{A'} \epsilon_{AC} \bar{\epsilon}_{C'B'} \tag{303}$$

The spinor notation discussed here (which is due to Penrose) essentially avoids the γ-matrices by choosing a basis for neither spinor space nor Minkowski space. Questions of (restricted) Lorentz invariance simply do not arise: one cannot, with this notation, write anything which is not invariant.

19. The Dirac Equation

The field which describes a free, massive, spin-$\frac{1}{2}$ particle consists of a pair, $(\xi^A, \eta_{A'})$, of spinor fields on Minkowski space. These fields must satisfy the *Dirac equation*:

$$\nabla_{AA'}\xi^A = \frac{\mu}{\sqrt{2}}\eta_{A'} \tag{304}$$

$$\nabla^{AA'}\eta_{A'} = -\frac{\mu}{\sqrt{2}}\mu\xi^A \tag{305}$$

where μ is a positive real number (which, as we shall see shortly, is essentially the mass of the particle.) The Dirac equations are to be considered as analogous to the Klein-Gordon equation, or to the Maxwell equations.

There is another, for some purposes more illuminating, form for Dirac's equations. Taking the derivative of (304), we have

$$\nabla^{BA'}\nabla_{AA'}\xi^A = \frac{\mu}{\sqrt{2}}\nabla^{BA'}\eta_{A'} \tag{306}$$

Substituting (305) on the right in (306), and using the fact that

$$\nabla^{BA'}\nabla_{AA'} = \frac{1}{2}\delta^B{}_A\Box \tag{307}$$

we obtain

$$(\Box + \mu^2)\xi^B = 0 \tag{308}$$

Thus, the Dirac equations imply that the spinor field ξ^A (and, by a similar argument, $\eta_{A'}$) satisfies the Klein-Gordon equation. Conversely, if ξ^A is any solution of (308), then, defining $\eta_{A'}$ by (304) (note $\mu \neq 0$), $(\xi^A, \eta_{A'})$ is a solution of Dirac's equations. In other words, there is a one-to-one correspondence between solutions of Dirac's equations and solutions of (308). Why, then, do we choose to deal with a pair of spinor fields and the relatively complicated equations (304), (305) rather than simply a simple spinor field and (308)? The reason is that there is a certain symmetry between ξ^A and $\eta_{A'}$ which, while merely a curiosity at present, will later be found to be related to the discrete symmetries of Minkowski space.

One further consequence of (308) is that it makes clear the fact that the problem of finding solutions of Dirac's equations is no more and no less difficult

83

than that of finding solutions of the Klein-Gordon equations. Fix two constant spinor fields, α^A and β^A, on Minkowski space. Then, by the remarks above, each solution, $(\xi^A, \eta_{A'})$, of Dirac's equations defines two solutions, $\xi^A \alpha_A$ and $\xi^A \beta_A$, of the Klein-Gordon equation, and, conversely, if ϕ and ψ are solutions of the Klein-Gordon equation, then $\phi \alpha^A + \psi \beta^A$ is a solution of (308), and hence defines a solution of Dirac's equations.

Note that if $(\xi^A, \eta_{A'})$ is a solution of Dirac's equations, so is $(\bar{\eta}^A, \bar{\xi}_{A'})$. We call $(\bar{\eta}^A, \bar{\xi}_{A'})$ the *complex-conjugate* of the solution $(\xi^A, \eta_{A'})$ (analogous to the complex-conjugate of a solution of the Klein-Gordon equation). Of course, complex-conjugation, applied twice to a solution of Dirac's equations, yields the original solution.

We now go to momentum space. Set

$$\xi^A(x) = \int_{M_\mu} \xi^A(k) e^{ik_b x^b} \, dV_\mu \tag{309}$$

$$\eta_{A'}(x) = \int_{M_\mu} \eta_{A'}(k) e^{ik_b x^b} \, dV_\mu \tag{310}$$

where $\xi^A(k)$ and $\eta_{A'}(k)$ are spinor-valued functions. These functions are only defined on M_μ, and the integrals (309), (310) are only carried out over M_μ, because of (308). Inserting (309) and (310) into (304) and (305), we obtain

$$ik_{AA'} \xi^A(k) = \frac{\mu}{\sqrt{2}} \eta_{A'}(k) \tag{311}$$

$$ik^{AA'} \eta_A(k) = -\frac{\mu}{\sqrt{2}} \xi^A(k) \tag{312}$$

Note that each of (311) and (312) implies the other. Thus, a solution of Dirac's equations is characterized by a pair of spinor-valued functions, $\xi^A(k)$ and $\eta_{A'}(k)$, on M_μ, subject to (311) and (312). (Alternatively, a solution is characterized by a single, arbitrary, spinor function $\xi^A(k)$ on M_μ. Then $\eta_{A'}(k)$ is defined by (311), and (312) follows identically.) A solution of Dirac's equations is said to be *positive-frequency* (resp. *negative-frequency*) if $\xi^A(k)$ and $\eta_{A'}(k)$ vanish on M_μ^- (resp. M_μ^+). In momentum space, complex conjugation has the effect

$$\xi^A(k) \to \bar{\eta}^A(-k) \qquad \eta_{A'}(k) \to \bar{\xi}_{A'}(-k) \tag{313}$$

Thus, just as in the Klein-Gordon case, complex-conjugation takes positive-frequency solutions to negative-frequency solutions, and vice-versa. (Roughly speaking, positive-frequency solutions represent electrons, and negative-frequency solutions positrons.)

Let $(\xi^A, \eta_{A'})$ be a solution of Dirac's equations, and consider the real vector field

$$j^a = \xi^A \bar{\xi}^{A'} + \eta^{A'} \bar{\eta}^A \tag{314}$$

in Minkowski space. First note that, since each term on the right in (314) is a future-directed null vector, j^a is future-directed and either timelike or null. We have, for the divergence of j^a,

$$\nabla_a j^a = \xi^A \nabla_{AA'} \bar{\xi}^{A'} + \bar{\xi}^{A'} \nabla_{AA'} \xi^A + \eta^{A'} \nabla_{AA'} \bar{\eta}^A + \bar{\eta}^A \nabla_{AA'} \eta^{A'} \tag{315}$$

Substituting (304) and (305), and using (292), we find

$$\nabla_a j^a = 0 \tag{316}$$

Thus, j^a is a real, future-directed timelike or null, divergence-free vector field. Therefore, the integral of j^a over a spacelike 3-plane yields a nonnegative number which, assuming that the Dirac field goes to zero sufficiently quickly at infinity, is independent of the choice of the 3-plane. This integral can be used to define a norm on solutions of Dirac's equations. The situation is much simpler when translated into momentum space (see (23)). We define the norm by

$$\frac{i\sqrt{2}}{\mu} \left(\int_{M_\mu^+} \xi^A(k) \bar{\eta}_A(k) \, dV_\mu - \int_{M_\mu^-} \xi^A(k) \bar{\eta}_A(k) \, dV_\mu \right) \tag{317}$$

Note that, because of (311) and (312), the expression (317) is equal to both of

$$\left(\int_{M_\mu^+} - \int_{M_\mu^-} \right) \frac{1}{\mu^2} (\xi^A \bar{\xi}^{A'} + \bar{\eta}^A \eta^{A'}) k_{AA'} \, dV_\mu \tag{318}$$

$$\left(\int_{M_\mu^+} - \int_{M_\mu^-} \right) \frac{2}{\mu^2} \xi^A \bar{\xi}^{A'} k_{AA'} \, dV_\mu \tag{319}$$

The forms (318) or (319) show, in particular, that our norm is positive. (The vector $\xi^A \bar{\xi}^{A'}$ is future-directed null, whereas $k_{AA'}$ is future-directed timelike on M_μ^+ and past-directed timelike on M_μ^-.)

We have now obtained a norm on our collection of solutions of Dirac's equation. In order to obtain a Hilbert space, therefore, we have only to impose the structure of a complex vector space on our collection of solutions. In other words, we must define addition of solutions and multiplication of solutions by complex numbers. There is only one reasonable way to define addition: one simply adds the corresponding spinor fields in Minkowski space (or, in momentum space, adds the corresponding spinor functions on the mass shell.) One might think, at first glance, that there is also only one reasonable definition of the product of a complex number and a solution of Dirac's equations: if α is a complex number, and $(\xi^A(k), \eta_{A'}(k))$ is a solution of Dirac's equations, one defines the product to be the solution $(\alpha \xi^A(k), \alpha \eta_{A'}(k))$. In other words, since the Dirac equation is linear on (complex) spinor fields, the solutions of this equation naturally have the structure of a complex vector space. There is, however, an alternative way to define the product of a solution of Dirac's equations and a complex number. Let $\xi(k)$ and $\eta_{A'}(k)$ be a pair of spinor functions on M_μ which satisfy (311) and (312), i.e., a solution (in momentum space) of Dirac's equations. Let α be a complex number. Then we might also define the product of α and $(\xi^A(k), \eta_{A'}(k))$ to be the solution

$$\begin{aligned} (\alpha \xi^A(k), \alpha \eta_{A'}(k)) & \qquad \text{for } k \in M_\mu^+ \\ (\bar{\alpha} \xi^A(k), \bar{\alpha} \eta_{A'}(k)) & \qquad \text{for } k \in M_\mu^- \end{aligned} \tag{320}$$

of Dirac's equations. That is to say, we multiply the positive-frequency part of the fields by α and the negative-frequency part by $\bar{\alpha}$. We obtain, in this way, an essentially different complex vector space of solutions of Dirac's equations. In fact, we adopt this second — rather less aesthetic — alternative. As we shall see later, this choice is essential to obtain agreement between theory and experiment.

We now have a complex vector space with a norm, (317), and hence a Hilbert space. More precisely, the Hilbert space of the Dirac equation, H_D, is the collection of all pairs, $(\xi^A(k), \eta_{A'}(k))$, of spinor functions on M_μ which satisfy (311), which are measurable, and for which the integral (317) converges. The inner product on our Hilbert space can now be obtained from the norm via the identity

$$(\phi, \psi) = \frac{1}{4} \left(\|\phi + \psi\|^2 - \|\phi - \psi\|^2 \right) + \frac{i}{4} \left(\|\psi + i\psi\|^2 - \|\phi - i\psi\|^2 \right) \qquad (321)$$

Using (317) and (321), the inner product on H_D takes the form

$$\frac{1}{\mu^2} \int_{M_\mu^+} \left(\xi^A \bar{\sigma}^{A'} + \eta^{A'} \bar{\tau}^A \right) k_{AA'} \, dV_\mu - \frac{1}{\mu^2} \int_{M_\mu^-} \left(\bar{\xi}^{A'} \sigma^A + \bar{\eta}^A \tau^{A'} \right) k_{AA'} \, dV_\mu$$
$$(322)$$

where $(\xi^A(k), \eta_{A'}(k))$ and $(\sigma^A(k), \tau_{A'}(k))$ are two solutions of Dirac's equations. Note the appearance of the complex-conjugations in the integral over M_μ^-. These arise because of our choice of the complex vector space structure for H_D.

To summarize, whereas the solutions of Dirac's equations have only one reasonable real vector space structure and only one reasonable norm, there are two possible complex vector space structures, of which we choose one. This choice then leads to the particular form for the inner product on our Hilbert space.

We now introduce the antisymmetric Fock space based on H_D. We thus have creation and annihilation operators, number operators, etc.

In the real Klein-Gordon and Maxwell cases, we were dealing with real fields on Minkowski space. This feature was reflected in momentum space by our requirement that the fields on the mass shell be invariant under simultaneous complex-conjugation and reflection through the origin. Physically, we were dealing with particles which are identical with their antiparticles. While we could, of course, restrict ourselves to real ($\xi^A = \bar{\eta}^A$) solutions of Dirac's equations, it is convenient not to do so. Thus, the functions on the future mass shell need bear no special relation to those on the past mass shell. This state of affairs leads to a pair of projection operators on H_D. Let $(\xi^A(k), \eta_{A'}(k)) \in H_D$. Then the action of P^+ (projection onto the positive frequency part) is defined by

$$P^+(\xi^A(k), \eta_{A'}(k)) = \begin{cases} (\xi^A, \eta_{A'}) & \text{for} \quad k \in M_\mu^+ \\ (0, 0) & \text{for} \quad k \in M_\mu^- \end{cases} \qquad (323)$$

and similarly for P^-. Note that

$$P^+ + P^- = \mathbb{I} \qquad (324)$$

These operators are both projection operators, i.e., they are defined everywhere and satisfy

$$(P^+)^2 = P^+ \qquad (P^-)^2 = P^- \qquad (325)$$

Eigenstates of P^+ with eigenvalue one (i.e., positive-frequency solutions) are called *particle states*, and those of P^- *antiparticle states*. Thus, we can speak of creation or annihilation of particle and antiparticle states, number operators for particles and antiparticles, etc. When we discuss charge, for example, we shall introduce the total charge operator, $eP^- - eP^+$, (this is the form when the particles have negative charge, e.g., electrons) where e is the fundamental charge.

The Dirac equation describes particles of mass $\hbar\mu$ and spin $\frac{1}{2}$. This statement must, of course, be proven using the techniques of Sect. 17. We now give the proof. The only piece of additional machinery we require is the notion of the Lie derivative of a spinor field. Quite generally, any smooth mapping, with smooth inverse, from Minkowski space to itself takes any tensor field on Minkowski space to another tensor field on Minkowski space. Smooth mappings which "differ infinitesimally from the identity mapping" are described by smooth vector fields. The corresponding "infinitesimal change in a tensor field" defines the Lie derivative of that tensor field. Does a smooth mapping, with smooth inverse, on Minkowski space take spinor fields to spinor fields? In other words, can we formulate a natural notion of the Lie derivative of a spinor field (by a vector field) so that the Lie derivative of a tensor field will arise as a special case (i.e., considering a tensor field as merely a special case of a spinor field when the numbers of primed and unprimed spinor indices are equal)? Unfortunately, the answer to these questions is no. To see this, suppose for a moment that it were possible to generalize the Lie derivative from tensor to spinor fields. Let v^a be an arbitrary smooth vector field on Minkowski space. Then we would have

$$\mathscr{L}_v \eta_{ab} = \mathscr{L}_v(\epsilon_{AB}\bar{\epsilon}_{A'B'}) = \epsilon_{AB}\mathscr{L}_v\bar{\epsilon}_{A'B'} + \bar{\epsilon}_{A'B'}\mathscr{L}_v\epsilon_{AB} \tag{326}$$

But, since ϵ_{AB} is skew, so must be $\mathscr{L}_v\epsilon_{AB}$, and similarly for $\mathscr{L}_v\bar{\epsilon}_{A'B'}$. Thus, the right side of (326) must be some multiple of the Minkowski metric η_{ab}. But it is simply false that, for an arbitrary smooth vector field v^a on Minkowski space, $\mathscr{L}_v\eta_{ab}$ is a multiple of η_{ab}. Thus, we cannot in general define the Lie derivative of a spinor field. Intuitively, the problem is that the light-cone structure of Minkowski space is an essential ingredient in the very definition of a spinor field. A smooth (finite or infinitesimal) mapping on Minkowski space which alters the light-cone structure simply does not know what to do with a general spinor field.

The remarks above are also the key to resolving the problem. In order to define spin and mass, it is only necessary to take Lie derivatives of spinor fields by vector fields v^a which satisfy (216) — i.e., by vector fields which do preserve the light-cone structure of Minkowski space. We might expect to be able to define Lie derivatives by such vector fields, and this is indeed the case. The formula is, for example,

$$\mathscr{L}_v T^{ABC'}{}_{DE'} = v^m \nabla_m T^{ABC'}{}_{DE'} - \frac{1}{2}T^{MBC'}{}_{DE'}\nabla_{MM'}v^{AM'}$$

$$- \frac{1}{2}T^{AMC'}{}_{DE'}\nabla_{MM'}v^{BM'} - \frac{1}{2}T^{ABM'}{}_{DE'}\nabla_{MM'}v^{MC}$$

$$+ \frac{1}{2}T^{ABC'}{}_{ME'}\nabla_{DM'}v^{MM'} + \frac{1}{2}T^{ABC'}{}_{DM'}\nabla_{ME'}v^{MM'} \tag{327}$$

Note that Lie differentiation commutes with complex-conjugation (v^a is real), raising and lowering of spinor indices, and contraction of spinor indices. Note, furthermore, that (327) reduces to the usual Lie derivative for tensor fields. It follows from the remarks on p. 80 that (327) is the only formula which satisfies these properties.

We first determine the mass associated with the Dirac equation. Let $(\xi^A(x), \eta_{A'}(x))$ be a solution of the Dirac equation, and r^a a vector at some point x of Minkowski space. Then

$$(r^a \underline{P}_a)\xi^M = \frac{\hbar}{i} r^a \nabla_a \xi^M$$
$$(r^a \underline{P}_a)(r^b \underline{P}_b)\xi^M = \hbar^2 r^a r^b \nabla_a \nabla_b \xi^M \tag{328}$$

Substituting η^{ab} for $r^a r^b$, we obtain

$$\underline{P}_a \underline{P}^a \xi^M = -\hbar^2 \Box \xi^M = \hbar^2 \mu^2 \xi^M \tag{329}$$

where we have used (308). Hence, from (268), the mass associated with the Dirac equation is $\hbar\mu$. The spin calculation is slightly more complicated. Let s^{cd} be a skew tensor at x. Then, from (327),

$$s^{cd} \underline{P}_{cd} \xi^M = \frac{\hbar}{i}\left(s^c{}_d x^d \nabla_c \xi^M - \frac{1}{2}\xi^N s^{MN'}{}_{NN'} \right) \tag{330}$$

If r^b is a vector at x, we have, therefore.

$$r^b \underline{P}_b s^{cd} \underline{P}_{cd} \xi^M = \hbar^2 r^b \nabla_b \left(s^c{}_d x^d \nabla_c \xi^M - \frac{1}{2}\xi^N s^{MN'}{}_{NN'} \right)$$
$$= \hbar^2 r^{BB'} s^{CC'DD'} \left(\epsilon_{BD}\bar{\epsilon}_{B'D'} \nabla_{CC'}\xi^M + \frac{1}{2}\bar{\epsilon}_{D'C'}\delta^M{}_C \nabla_{BB'}\xi_D \right) \tag{331}$$

Substituting $u_a \epsilon^a bcd$ for $r^b s^{cd}$, and using (269) and (302),

$$u^a \underline{W}_a \xi^M = i\hbar^2 \left(\epsilon^{AB}\epsilon^{CD}\bar{\epsilon}^{A'C'}\bar{\epsilon}^{B'D'} - \bar{\epsilon}^{A'B'}\bar{\epsilon}^{C'D'}\epsilon^{AC}\epsilon^{BD} \right) u_{AA'}$$
$$\times \left(\epsilon_{BD}\bar{\epsilon}_{B'D'}\nabla_{CC'}\xi^M + \frac{1}{2}\bar{\epsilon}_{D'C'}\delta^M{}_C \nabla_{BB'}\xi_D \right) \tag{332}$$
$$= -\frac{i}{2}\hbar^2 u^a \nabla_a \xi^M + i\hbar^2 u^{MA'} \nabla_{AA'}\xi^A$$

Therefore,

$$u^a \underline{W}_a u^b \underline{W}_b \xi^M = (i\hbar^2)^2 \left(\frac{1}{4}u^a u^b \nabla_a \nabla_b \xi^M - \frac{1}{2}u^{MA'}u^b \nabla_{AA'}\nabla_b \xi^A \right.$$
$$\left. -\frac{1}{2}u^{MB'}u^a \nabla_{BB'}\nabla_a \xi^B + u^{MA'}u^{AB'}\nabla_{AA'}\nabla_{BB'}\xi^B \right) \tag{333}$$

Finally, substituting η^{ab} for $u^a u^b$, we have

$$\underline{W}^a \underline{W}_a \xi^M = \frac{3}{4}\hbar^4 \Box \xi^M = -\frac{3}{4}\hbar^2 m^2 \xi^M \tag{334}$$

We conclude from (270) that the Dirac equation describes a particle with spin $\frac{1}{2}$.

20. The Neutrino Equation

A neutrino is essentially a "massless Dirac particle". There are, however, a few features which are particular to the case $\mu = 0$.

The (four-component) neutrino field consists of a pair, $(\xi^A, \eta_{A'})$ of spinor fields on Minkowski space, subject to the *neutrino equation* (see (304), (305)):

$$\nabla_{AA'}\xi^A = 0 \tag{335}$$

$$\nabla^{AA'}\eta_{A'} = 0 \tag{336}$$

Note that, whereas in the massive case either of the two spinor fields can be obtained from the other (via (304), (305)), the fields become "uncoupled" in the massless case. That is to say, each spinor field satisfies its own equation. Taking a derivative of (335),

$$\nabla^{BA'}\nabla_{AA'}\xi^A = 0 \tag{337}$$

and using (307), we obtain

$$\Box\xi^A = 0 \tag{338}$$

and similarly for $\eta_{A'}$. Thus, each of our neutrino fields satisfies the wave equation. Note, however, that (338) does not imply (335). (Solutions of the neutrino equations can, however, be obtained from solutions of the wave equation. If $\alpha_{A'}$ satisfies the wave equation, then $\xi^A = \nabla^{AA'}\alpha_{A'}$ satisfies (335).)

The *complex-conjugate* of the solution $(\xi^A, \eta_{A'})$ of the neutrino equation is the solution $(\bar{\eta}^A, \bar{\xi}_{A'})$.

Passing to momentum space, we set

$$\xi^A(x) = \int_{M_0} \xi^A(k)e^{ik_b x^b}\,dV_0 \tag{339}$$

$$\eta_{A'}(x) = \int_{M_0} \eta_{A'}(k)e^{ik_b x^b}\,dV_0 \tag{340}$$

where $\xi^A(k)$ and $\eta_{A'}(k)$ are spinor-valued functions on the zero-mass shell, M_0. In momentum space, (335) and (336) become

$$\xi^A(k)k_{AA'} = 0 \tag{341}$$

$$\eta_{A'}(k)k^{AA'} = 0 \tag{342}$$

Positive-frequency and negative-frequency solutions of the neutrino equations are well-defined. Complex-conjugation again reverses frequency, and is again expressed in momentum apace by the equations (313).

The current (314) is still divergence-free in the massless case. (In fact, the proof is rather simpler with $\mu = 0$.) This fact leads to a norm on solutions of the neutrino equation. The simplest way to obtain the norm, however, is as a "$\mu \to 0$ limit" of the Dirac norm. Consider (318). It is not difficult to check from (311) and (312) that

$$\xi^A(k)\bar{\xi}^{A'}(k) + \eta^{A'}\bar{\eta}^A(k) = \alpha(k)k^{AA'} \tag{343}$$

where $\alpha(k)$ is a real function on M_μ which is positive on M_μ^+ and negative on M_μ^-. From (318), the norm in the Dirac case is simply

$$\int_{M_0} |\alpha(k)| \, dV_\mu \tag{344}$$

We now return to the massless case. Eqn. (341) implies that $\xi^A(k)\bar{\xi}^{A'}(k)$ is proportional to $k^{AA'}$, while (342) implies that $\eta^{A'}(k)\bar{\eta}^A(k)$ is also proportional to $k_{AA'}$. Therefore,

$$\xi^A(k)\bar{\xi}^{A'}(k) + \eta^{A'}\bar{\eta}^A(k) = \alpha(k)k^{AA'} \tag{345}$$

on M_0, where $\alpha(k)$ is real on M_0 and positive on M_0^+ and negative on M_0^-. We therefore define the norm in the neutrino case, in analogy with (344), by

$$\int_{M_0} |\alpha(k)| \, dV_\mu \tag{346}$$

For the complex vector space structure in the massless case, we use the same convention as in the massive case (see (320)).

In fact, the theory we have been discussing is not very interesting physically. The reason is that our Hilbert space of solutions of the neutrino equation contains four irreducible subspaces: positive-frequency solutions with $\eta_{A'} = 0$, negative-frequency solutions with $\eta_{A'} = 0$, positive-frequency solutions with $\xi^A = 0$, and negative-frequency solutions with $\xi^A = 0$. Every solution can be written uniquely as the sum of four solutions, one from each class above. Thus, our neutrino field describes four similar particles. But neutrinos in the real world appear in pairs (particle-antiparticle.) Thus, we would like to introduce a field whose Hilbert space has only two irreducible (under the restricted Poincaré group) subspaces. The result is what is called the "two-component neutrino theory", which we now describe. (The only purpose in treating the four-component theory at all was to make explicit the analogy with the Dirac equation.)

The (two-component) neutrino field is a single spinor field ξ^A on Minkowski space which satisfies (335), and, therefore, (338). In momentum space, we have a spinor-valued function $\xi^A(k)$ on M_0 which satisfies (341). This equation implies

$$\xi^A(k)\bar{\xi}^{A'}(k) = \alpha(k)k^{AA'} \tag{347}$$

where $\alpha(k)$ is real, and positive on M_0^+ and negative on M_0^-. We define the norm on our solutions by

$$\int_{M_0} |\alpha(k)| \, dV_0 \tag{348}$$

The complex vector space structure is defined as before: the product of a complex number β and a solution $\xi^A(k)$ is defined to be the solution

$$\begin{aligned} \beta \xi^A(k) &\quad \text{for } k \in M_0^+ \\ \bar{\beta} \xi^A(k) &\quad \text{for } k \in M_0^- \end{aligned} \tag{349}$$

The collection of all measurable spinor functions $\xi^A(k)$ on M_0 for which (341) is satisfied and (348) converges, with the above complex vector space structure, is a Hilbert space which we write as H_N.

We introduce the antisymmetric Fock space based on H_N. We thus have creation and annihilation operators, number operators, etc.

We introduce on H_N the two projection operators P^+ and P^-, projection onto positive and negative frequency, respectively. These operators, of course, satisfy (324) and (325).

Finally, we remark on the spin and mass to be associated with H_N. We have done the mass calculation several times: (338) clearly leads to $m = 0$ for H_N. Furthermore, most of the work involved in calculating the spin has already been done. Nowhere in the argument leading to (332) did we use the fact that ξ^A satisfies the Dirac equation, and so (332) holds also in the neutrino case. But now (335) implies that the second term on the right in (332) vanishes, so we have

$$u^a \underline{W}_a \xi^B = \frac{1}{2} \frac{\hbar^2}{i} u^a \nabla_a \xi^B \tag{350}$$

Since, furthermore,

$$u^a \underline{P}_a \xi^B = \frac{\hbar}{i} u^a \nabla_a \xi^B \tag{351}$$

one is tempted to conclude from (271) that $s = \frac{1}{2}$ for H_N. This conclusion is essentially correct, but one technical point must be clarified. (Unfortunately, our notation is rather badly suited to the remarks below, and so they will sound rather mystical.) The problem involves what the i's mean in (350) and (351). (This problem never arose in the Dirac case because the i's were always squared away, so their meaning was irrelevant.) Where did the i's come from? The i in (351) came from the $\hbar/i$ factors which are introduced in the operator fields associated with the Poincaré group. This i means "multiplication by i within the Hilbert space H_N" because only in this way does one obtain Hermitian operators from "infinitesimal unitary operators". In other words, the i in (351) arises from very general considerations involving the action of a Lie group on a Hilbert space, and, in this general framework within which the formalism was set up, there is only one notion of multiplication by i, namely, multiplication within the Hilbert space. Thus, the "i" in (351) multiplies the positive-frequency part of what follows by i, and the negative-frequency part by $-i$. (See (349).) The i in (350), on the other hand, is a quite different animal. It arose from the i in

(302). (The i's in $\underline{P}_a$ and $\underline{P}_{ab}$ (see (269)) combine to give -1.) But the i in (302) appears because of the way that the real tensor field ϵ^{abcd} must be expressed in terms of spinors. Hence, the "i" in (350), because of its origin, represents simply multiplication of a tensor field by i. That is to say, the "i-operators" in (350) and (351) are equal for positive-frequency solutions, and minus each other for negative-frequency solutions. Thus, $s = \frac{1}{2}$ for positive-frequency solutions (neutrinos), and $s = -\frac{1}{2}$ for negative-frequency solutions (antineutrinos). That is, in the neutrino case the particle and its antiparticle have opposite helicity. This "prediction" is in fact confirmed by experiment.

21. Complex Klein-Gordon Fields

In Sect. 5, we dealt with real equations of the Klein-Gordon equation (although, for reasons of motivation, we chose to characterize such fields as complex positive-frequency solutions). Such fields describe particles with spin zero which are identical with their antiparticles (e.g., the π^0). On the other hand, there are spin-zero particles which are not identical with their antiparticles (the π^+ and π^-). These are described by complex solutions of the Klein-Gordon equation.

Consider a complex scalar field, $\phi(x)$, in Minkowski space which satisfies the Klein-Gordon equation, (5). In momentum space,

$$\phi(x) = \int_{M_\mu} \phi(k)e^{ik_b x^b}\, \mathrm{d}V_\mu \tag{352}$$

Thus, our solution is characterized by a complex-valued function $\phi(k)$ on M_μ. (In the real case, one requires in addition $\phi(-k) = \bar{\phi}(k)$.) The norm of such a function is defined by

$$\frac{1}{\hbar} \int_{M_\mu} \phi(k)\bar{\phi}(k)\, \mathrm{d}V_\mu \tag{353}$$

We adopt, for the complex vector space structure on these functions, essentially the same structure used in the Dirac and neutrino case. To "multiply" $\phi(k)$ by a complex number, one takes

$$\begin{array}{ll} \alpha\phi(k) & \text{for } k \in M_\mu^+ \\ \bar{\alpha}\phi(k) & \text{for } k \in M_\mu^- \end{array} \tag{354}$$

The collection of all measurable, square-integrable (in the sense of (353), complex-valued functions on M_μ, with this complex vector space structure, is our Hilbert space, H_{CKG}, for complex solutions of the Klein-Gordon equation. (We shall write the Hilbert space of Sect. 5 as H_{RKG}.)

There is defined on H_{CKG} the two projection operators, P^+ and P^-, which take the positive-frequency part and negative-frequency part, respectively.

We introduce the symmetric Fock space based on H_{CKG}, creation and annihilation operators, number operators, etc.

22. Positive Energy

Many of the quantities associated with an elementary particle (e.g., charge) are reversed in the passage from a particle to its antiparticle. It is observed experimentally, however, that energy is not one of these quantities. For example, if an electron and a positron annihilate (say, with negligible kinetic energy), then the total, energy released is $2m$, and not zero. We are thus forced to assign a (rest) energy $+m$ to both a positron and an electron. Where does this fact appear in our formalism?

Of course, "energy" refers to the state of motion of an observer. This "state of motion" is represented by some constant, unit, future-directed timelike vector field r^a in Minkowski space. The energy operator is then $\underline{E} = r^a \underline{P}_a$. It should be emphasized that we are not free to assign energies arbitrarily to obtain agreement with experiment. The very concept of energy is based in an essential way on the action of the Poincaré group (more explicitly, on the time translations). If we wish to avoid a radical change in what energy means in the passage from classical to quantum theory, we must choose for the energy in quantum field theory that quantity which arises naturally from time translations in Minkowski space, i.e., we must choose the $\underline{E}$ above. We take as our precise statement that "energies are nonnegative" the statement that the expectation value of $\underline{E}$ in any state σ (on which $\underline{E}$ is defined) be nonnegative:

$$(\sigma, \underline{E}\sigma) \geq 0 \qquad (355)$$

Is it true or false that (355) holds for the five Hilbert spaces we have constructed, H_{RKG}, H_{CKG}, H_M, H_D, H_N?

We begin with the real Klein-Gordon case. The Hilbert space consists of measurable, square-integrable, complex-valued functions $\phi(k)$ on M_μ which satisfy

$$\phi(-k) = \bar{\phi}(k) \qquad (356)$$

Such functions do not have an obvious complex vector space structure. If $\phi(k)$ satisfies (356), and α is a complex number, then $\alpha\phi(k)$ will not in general satisfy (356). This fact, of course, is not surprising: there is no obvious way to take the "product" of a complex number and a real solution of a differential equation to obtain another real solution. This problem is resolved, in H_{RKG}, by choosing one of the two mass shells to be preferred, and calling it the "future" mass shell, M_μ^+. We then agree that, to multiply by α, "M_μ^+ gets α while M_μ^- must be content with $\bar{\alpha}$." In other words, we define multiplication of $\phi(k)$ by α by

$$
\begin{array}{ll}
\alpha\phi(k) & k \in M_\mu^+ \\
\bar{\alpha}\phi(k) & k \in M_\mu^-
\end{array}
\tag{357}
$$

It should be emphasized that, in the real case, we are forced (by the requirement that we obtain a Hilbert space) to select one preferred mass shell and define multiplication by (357).

Now consider the energy. If $\phi(x)$ is a real solution of the Klein-Gordon equation, then

$$
\underline{E}\phi(x) = \frac{\hbar}{i}r^a\nabla_a\phi(x)
\tag{358}
$$

Because of the i in (358), one might naively think that (358) does not represent a real solution of the Klein-Gordon equation, and so that (358) is not a definition for $\underline{E}$. This, of course, is not the case. The i in (358) arose because of general considerations involving representations of the Poincaré group (Sect. 16), and means "multiplication within the Hilbert space H_{RKG}." In momentum space, the operator $\hbar r^a\nabla_a$ has the effect

$$
\phi(k) \to i\hbar(r^a k_a)\phi(k)
\tag{359}
$$

Note that (359) does not destroy (356), a statement which reflects the fact that

$$
\phi(x) \to \hbar r^a\nabla_a\phi
\tag{360}
$$

is an unambiguous operation on real solutions of the Klein-Gordon equation. Now using (357), the energy operator in momentum space takes the form

$$
\phi(k) \to \begin{cases} \hbar r^a k_a\phi(k) & k \in M_\mu^+ \\ -\hbar r^a k_a\phi(k) & k \in M_\mu^- \end{cases}
\tag{361}
$$

The expectation value of $\underline{E}$ in the state $\phi(k)$ is

$$
\int_{M_\mu^+} (r^a k_a)\phi(k)\bar{\phi}(k)\,dV_\mu - \int_{M_\mu^-} (r^a k_a)\phi(k)\bar{\phi}(k)\,dV_\mu
\tag{362}
$$

which, of course, is positive. (Why don't we just define the energy operator by (359), (360), leaving out the i? Because the expectation value of this operator is not real. That is, the i is needed for Hermiticity.)

We summarize the situation. In order to make a Hilbert space of real solutions of the Klein-Gordon equation, we are forced to select a preferred mass shell to be called "future". Then, provided r^a is "future-directed" according to this convention, $\underline{E}$ will have positive expectation values.

Now consider the complex Klein-Gordon case. The energy operator still has the form

$$
\underline{E}\phi(x) = \frac{\hbar}{i}r^a\nabla_a\phi(x)
\tag{363}
$$

and i still means multiplication within our Hilbert space. In momentum space, $\phi(k)$ is an arbitrary measurable, square-integrable, complex-valued function. The operator $\hbar r^a \nabla_a$ has the effect

$$\phi(k) \to i\hbar(r^a k_a)\phi(k) \tag{364}$$

We must still multiply (364) by $1/i$. But we now have the freedom to select one of two possible complex vector space structures on the complex solutions of the Klein-Gordon equation. For the "product" of a complex number α and $\phi(k)$, we could choose

$$\alpha\phi(k) \tag{365}$$

or, alternatively,

$$\begin{aligned} \alpha\phi(k) \qquad & k \in M_\mu^+ \\ \bar{\alpha}\phi(k) \qquad & k \in M_\mu^- \end{aligned} \tag{366}$$

The resulting energy operators are

$$\underline{E}\phi(k) = \hbar(r^a k_a)\phi(k) \tag{367}$$

$$\underline{E}\phi(k) = \begin{cases} \hbar r^a k_a \phi(k) & k \in M_\mu^+ \\ -\hbar r^a k_a \phi(k) & k \in M_\mu^- \end{cases} \tag{368}$$

respectively. Finally, the resulting expectation values of $\underline{E}$ are

$$\int_{M_\mu} r^a k_a \phi(k)\bar{\phi}(k)\, dV_\mu \tag{369}$$

$$\int_{M_\mu^+} (r^a k_a)\phi(k)\bar{\phi}(k)\, dV_\mu - \int_{M_\mu^-} (r^a k_a)\phi(k)\bar{\phi}(k)\, dV_\mu \tag{370}$$

respectively. But note that (369) can take both positive and negative values, while (370) is always nonnegative. But this is exactly what one might expect. The complex vector space structure (365) does not prefer one time direction over the other: it makes no reference to past and future. Therefore, it could not possibly lead to a positive energy, for the energy associated with r^a is certainly minus the energy associated with $-r^a$. The complex vector space structure (366), on the other hand, picks out a particular "future" time direction. Then the expectation value of $\underline{E}$ is positive provided r^a is "future-directed" in this sense. It is for this reason that we are led to select (366) as our complex vector space structure.

We summarize. If energy is to arise from time translations, there is no freedom to alter the energy operator itself. In the real Klein-Gordon case, we are forced, in order to obtain a Hilbert space, to select a preferred "future" mass shell. Then energy is positive provided r^a is future-directed. In the complex Klein-Gordon case, there are two distinct ways to obtain a Hilbert space, one which selects a preferred "future" mass shell, and one which does not. It is only the former choice which leads to positive energies. We make this choice.

There is an additional sense in which (366) is a more natural choice for the complex vector space structure for $H_C KG$. Every real solution of the Klein-Gordon equation is certainly also a complex solution. We thus have a mapping $\Lambda : H_{RKG} \to H_{CKG}$. This mapping is certainly norm-preserving. Is it also linear? The answer is no if we choose the structure (365), and yes if we choose the the the structure (366).

A completely analogous situation holds for the other Hilbert spaces. H_M is based on real solutions of Maxwell's equations, its complex vector space structure depends on choosing a particular future mass shell, and energies are naturally positive. On the other hand, H_D and H_N are based on complex fields. We have two choices for the complex vector space structure, one of which leads to positive energies and one of which does not. We choose the complex vector space structure to be the one which, by preferring a future mass shell, makes the energy be positive. (See (320), (349).)

23. Fields as Operators: Propagators

Ordinary relativistic fields are to be replaced, eventually, by an appropriate class of operators on Fock space. This transition from fields to operators is to be carried out according to the following general rules:

i) A real field becomes a Hermitian operator; a pair of complex-conjugate fields is a pair of adjoint operators;

ii) The operators have the same index structure, and satisfy the same equations, as the corresponding fields; and

iii) The "positive-frequency part" of the operator is annihilation of a particle, the negative-frequency part creation of an antiparticle.

We have already discussed these operators in the real Klein-Gordon and Maxwell cases (Sects. 12 and 14, respectively). The purposes of this section are, firstly, to treat the complex Klein-Gordon and Dirac cases, and, secondly, to establish certain properties of the functions which appear in the commutators or anti-commutators. For completeness, we briefly review Sects. 12 and 14.

Real Klein-Gordon. H_{RKG} consists of (measurable, square-integrable) complex-valued functions $\phi(k)$ on M_μ which satisfy $\bar{\phi}(k) = \phi(-k)$. The inner product is

$$(\phi(k), \psi(k)) = \frac{1}{\hbar} \int_{M_\mu^+} \phi(k)\overline{\psi(k)} \, dV_\mu + \frac{1}{\hbar} \int_{M_\mu^-} \overline{\phi(k)}\psi(k) \, dV_\mu \tag{371}$$

Let $f(x)$ be a real test function on Minkowski space, and let $f(k)$ be its Fourier inverse, so $\bar{f}(k) = f(-k)$. Then $f(k)$, restricted to M_μ, defines an element, $\sigma(f)$, of H_{RKG}. The corresponding field operator on symmetric Fock space is

$$\underline{\phi}(f) = \hbar C(\sigma(f)) + \hbar A(\sigma(f)) \tag{372}$$

Note that (372) is Hermitian and satisfies the Klein-Gordon equation:

$$\underline{\phi}\left((\Box + \mu^2)f\right) = 0 \tag{373}$$

99

The commutator is

$$
\begin{aligned}
\left[\underline{\phi}(f), \underline{\phi}(g)\right] &= \hbar^2 \left([C(\sigma(f)), A(\sigma(g))] + [A(\sigma(f)), C(\sigma(g))]\right) \\
&= \hbar^2 \left(-\sigma^\alpha(f)\bar{\sigma}_\alpha(g) + \sigma^\alpha(g)\bar{\sigma}_\alpha(f)\right) \mathbb{I} \\
&= \frac{\hbar}{i} D(f,g)\mathbb{I}
\end{aligned}
\tag{374}
$$

where we have defined

$$
D(f,g) = -i \left(\int_{M_\mu^+} - \int_{M_\mu^-}\right) \left(f(k)\bar{g}(k) - \bar{f}(k)g(k)\right) \, dV_\mu
\tag{375}
$$

Complex Klein-Gordon. H_{CKG} consists of (measurable, square-integrable) complex-valued functions $\phi(k)$ on M_μ. The inner product is

$$
(\phi(k), \psi(k)) = \frac{1}{\hbar} \int_{M_\mu^+} \phi(k)\overline{\psi(k)} \, dV_\mu + \frac{1}{\hbar} \int_{M_\mu^-} \overline{\phi(k)}\psi(k) \, dV_\mu
\tag{376}
$$

Let $f(x)$ be a real test function on Minkowski space, and let $f(k)$ be its Fourier inverse, so $\bar{f}(k) = f(-k)$. Let $\sigma^+(f)$ be the element of H_{CKG} given by $f(k)$ on M_μ^+ and zero on M_μ^-, and let $\sigma^-(f)$ be given by $f(k)$ on M_μ^- and zero on M_μ^+. The corresponding field operators on symmetric Fock space are

$$
\underline{\phi}(f) = \hbar \left(C(\sigma^-(f)) + A(\overline{\sigma^+(f)})\right)
\tag{377}
$$

$$
\underline{\phi}^*(f) = \hbar \left(A(\overline{\sigma^-(f)}) + C(\sigma^+(f))\right)
\tag{378}
$$

Note that these are adjoints of each other, and that they satisfy the Klein-Gordon equation:

$$
\underline{\phi}\left((\Box + \mu^2)f\right) = \underline{\phi}^*\left((\Box + \mu^2)f\right) = 0
\tag{379}
$$

We clearly have

$$
\left[\underline{\phi}(f), \underline{\phi}(g)\right] = \left[\underline{\phi}^*(f), \underline{\phi}^*(g)\right]
\tag{380}
$$

For the other commutator, however,

$$
\begin{aligned}
\left[\underline{\phi}(f), \underline{\phi}^*(g)\right] &= \hbar^2 \left(\left[C(\sigma^-(f)), A(\overline{\sigma^-(g)})\right] + \left[A(\overline{\sigma^+(f)}), C(\sigma^+(g))\right]\right) \\
&= \hbar^2 \left(-\sigma^{-\alpha}(f)\overline{\sigma_\alpha^-}(g) + \sigma^{+\alpha}\overline{\sigma_\alpha^+}(f)\right) \mathbb{I} = \frac{\hbar}{2i} D(f,g)\mathbb{I}
\end{aligned}
\tag{381}
$$

where $D(f,g)$ is given by (375).

Maxwell. H_M consists of (measurable, square-integrable) complex vector functions $A_a(k)$ on M_0 which satisfy $\bar{A}_a(k) = A_a(-k)$ and $k^a A_a(k) = 0$, where two such functions which differ by a multiple of k_a are to be regarded as defining the same element of H_M. The inner product is

$$
(A_a(k), B_a(k)) = -\frac{1}{\hbar} \int_{M_0^+} A_a(k)\bar{B}^a(k) \, dV - \frac{1}{\hbar} \int_{M_0^-} \bar{A}_a(k)B^a(k) \, dV
\tag{382}
$$

Let $f^a(x)$ be a real test vector field on Minkowski space which is the sum of a divergence-free field and a gradient, and let $f^a(k)$ be its Fourier inverse. Then $f^a(k)$ satisfies $f^a(-k) = \bar{f}^a(k)$ and $k_a f^a(k) = 0$, and so defines an element, $\sigma(f^a)$, of H_M. The corresponding operator on symmetric Fock space is

$$\underline{A}(f^a) = \hbar C(\sigma(f^a)) + \hbar A(\sigma(f^a)) \tag{383}$$

This operator is Hermitian, and satisfies Maxwell's equations (for the vector potential):

$$\underline{A}(\nabla^a f) = 0 \qquad \underline{A}(\Box f^a) = 0 \tag{384}$$

The commutator is

$$[\underline{A}(f^a), \underline{A}(g^a)] = \hbar^2 \left(-\sigma^\alpha(f^a)\bar{\sigma}_\alpha(g^a) + \sigma^\alpha(g^a)\bar{\sigma}_\alpha(f^a) \right) \mathbb{I} = \frac{\hbar}{i} D(f^a, g^a)\mathbb{I} \tag{385}$$

where we have defined

$$D(f^a, g^a) = i \left(\int_{M_0^+} - \int_{M_0^-} \right) \left(f_a(k)\bar{g}^a(k) - g_a(k)\bar{f}^a(k) \right) \, dV \tag{386}$$

Dirac. H_D consists of (measurable, square-integrable) pairs, $(\xi^A(k), \eta_{A'}(k))$, of spinor functions on M_μ which satisfy

$$ik_{AA'}\xi^A(k) = \frac{\mu}{\sqrt{2}}\eta_{A'}(k) \qquad ik^{AA'}\eta_{A'}(k) = -\frac{\mu}{\sqrt{2}}\xi^A(k) \tag{387}$$

The inner product is

$$((\xi^A, \eta_{A'}), (\sigma^A, \tau_{A'})) = \frac{2}{\mu^2} \int_{M_\mu^+} \xi^A(k)\bar{\sigma}^{A'} k_{AA'} \, dV$$
$$- \frac{2}{\mu^2} \int_{M_\mu^-} \sigma^A(k)\bar{\xi}^{A'} k_{AA'} \, dV \tag{388}$$

Let $f^A(x), \bar{f}_{A'}(x)$ be a pair of test spinor fields on Minkowski space which is real in the sense that the second field is the complex-conjugate of the first. (Of course, "test" means having compact support.) Let $f^A(k)$ be the Fourier inverse of $f^A(x)$, and consider the pair

$$\left(\mu f^A(k) - \frac{i\sqrt{2}}{\mu} k^{AA'} \bar{f}_{A'}(k), \mu \bar{f}_{A'}(k) + \frac{i\sqrt{2}}{\mu} k_{AA'} f^A(k) \right) \tag{389}$$

of spinor functions on M_μ. Note that the pair (389) satisfies (387). Let $\sigma^+(f^A, \bar{f}_{A'})$ be the element of H_D which is given by (389) on M_μ^+ and zero on M_μ^-, and let $\sigma^-(f^A, \bar{f}_{A'})$ be the element given by (389) on M_μ^- and zero on M_μ^+. The corresponding operators on antisymmetric Fock space are

$$\underline{\psi}(f^A, \bar{f}_{A'}) = C(\sigma^-(f^A, \bar{f}_{A'})) + A(\overline{\sigma^+(f^A, \bar{f}_{A'})}) \tag{390}$$
$$\underline{\psi}^*(f^A, \bar{f}_{A'}) = A(\overline{\sigma^-(f^A, \bar{f}_{A'})}) + C(\sigma^+(f^A, \bar{f}_{A'})) \tag{391}$$

These operators are adjoints of each other, and satisfy the Dirac equation:

$$\underline{\psi}\left(\mu f^A(k) - \frac{\sqrt{2}}{\mu}\nabla^{AA'}\bar{f}_{A'}(k), \mu\bar{f}_{A'}(k) + \frac{\sqrt{2}}{\mu}\nabla_{AA'}f^A(k)\right)$$

$$= \underline{\psi}^*\left(\mu f^A(k) - \frac{\sqrt{2}}{\mu}\nabla^{AA'}\bar{f}_{A'}(k), \mu\bar{f}_{A'}(k) + \frac{\sqrt{2}}{\mu}\nabla_{AA'}f^A(k)\right) = 0 \quad (392)$$

We clearly have

$$\{\underline{\psi}(f^A, \bar{f}_{A'}), \underline{\psi}(g^A, \bar{g}_{A'})\} = \{\underline{\psi}^*(f^A, \bar{f}_{A'}), \underline{\psi}^*(g^A, \bar{g}_{A'})\} = 0 \quad (393)$$

For the other anticommutator,

$$\{\underline{\psi}(f^A, \bar{f}_{A'}), \underline{\psi}^*(g^A, \bar{g}_{A'})\}$$
$$= \left(\sigma^{-\alpha}(f^A, \bar{f}_{A'})\bar{\sigma}_\alpha^-(g^A, \bar{g}_{A'}) + \sigma^{+\alpha}(g^A, \bar{g}_{A'})\bar{\sigma}_\alpha^+(f^A, \bar{f}_{A'})\right)\mathbb{I} \quad (394)$$
$$= D\left((f^A, \bar{f}_{A'}), (g^A, \bar{g}_{A'})\right)\mathbb{I}$$

where, using (388) and (389),

$$D\left((f^A, \bar{f}_{A'}), (g^A, \bar{g}_{A'})\right)$$
$$= 2\left(\int_{M_\mu^+} - \int_{M_\mu^-}\right)\left[\left(f^A(k)\bar{g}^{A'}(k) + \bar{f}^{A'}(k)g^A(k)\right)k_{AA'}\right.$$
$$\left. + \frac{i\mu}{\sqrt{2}}\left(f^A(k)g_A(-k) - \bar{f}^{A'}(k)\bar{g}_{A'}(-k)\right)\right]dV \quad (395)$$

The functions (375), (386), and (395) play a very special role in relativistic quantum field theory. They are called *Feynman propagators*. Several properties of the propagators follow immediately from the definitions. In the first place, they are all real. Secondly, we have the symmetries

$$D(f, g) = -D(g, f) \quad (396)$$

$$D(f^a, g^a) = -D(g^a, f^a) \quad (397)$$

$$D\left((f^A, \bar{f}_{A'}), (g^A, \bar{g}_{A'})\right) = D\left((g^A, \bar{g}_{A'}), (f^A, \bar{f}_{A'})\right) \quad (398)$$

Furthermore, since the propagators arise from commutators or anticommutators of the field operators, they satisfy the appropriate field equations:

$$D\left((\Box + \mu^2)f, g\right) = 0 \quad (399)$$

$$D(\nabla^a f, g^a) = 0 \qquad D(\Box f^a, g^a) = 0 \quad (400)$$

$$D\left(\left(\mu f^A(k) - \frac{\sqrt{2}}{\mu}\nabla^{AA'}\bar{f}_{A'}(k), \mu\bar{f}_{A'}(k)\right.\right.$$
$$\left.\left. + \frac{\sqrt{2}}{\mu}\nabla_{AA'}f^A(k)\right), (g^A, \bar{g}_{A'})\right) = 0 \quad (401)$$

Note also that the propagators are linear in the real test fields on which they depend.

A more remarkable property of the propagators is that they can all be expressed directly in terms of the Klein-Gordon propagator, $D(f, g)$. Let v^a and w^a be constant vector fields on Minkowski space, f and g real test functions, and consider the expression

$$- v^a w_a D(f, g) \tag{402}$$

Inserting (375) in (402), we see that (402) is precisely the right side of (386), provided we set $f^a = f v^a$, $g^a = w g^a$. Thus, we may define a function $D(f^a, g^a)$ for test fields of the form $f^a = f v^a$, $g^a = w g^a$ by (402). Then, assuming linearity in f^a and g^a, we extend the range of $D(f^a, g^a)$ to arbitrary real test vector fields. Finally, restricting f^a and g^a to fields which can be expressed as the sum of a divergence-free field and a gradient, we obtain precisely the Maxwell propagator.

The situation is similar in the Dirac case. Let ξ^A and σ^A be constant spinor fields, f and g real test functions. Then, from (395) and (375), we have

$$
\begin{aligned}
D\left((f^A, \bar{f}_{A'}), (g^A, \bar{g}_{A'})\right) \\
= 2D\left((\xi^A \bar{\sigma}^{A'} + \sigma^A \bar{\xi}^{A'}) \nabla_{AA'} f, g\right) + \frac{\mu}{\sqrt{2}} (\xi^A \sigma_A + \bar{\xi}^{A'} \bar{\sigma}_{A'}) D(f, g) \quad (403)
\end{aligned}
$$

But, again by linearity, if we know the Dirac propagator for test fields of the form $f(\xi^A, \bar{\xi}_{A'})$ with constant ξ^A, we know the Dirac propagator for all test fields. In this sense, then, the Dirac propagator follows already from the Klein-Gordon propagator.

We may now derive a particularly important property of the Feynman propagators. A function of a pair of test fields will be called *causal* if it vanishes whenever the test fields have relatively spacelike supports (see p. 48). We have seen in Sect. 12 that the Klein-Gordon propagator, $D(f, g)$, is causal. The remarks above imply, therefore, that all the Feynman propagators are causal.

24. Spin and Statistics

What is it that determines whether the appropriate Fock space for an elementary particle is the symmetric or the antisymmetric one? (This distinction is said to one of *statistics*. Particles described by the symmetric Fock space are called *bosons*, and are said to satisfy *Bose statistics*. Particles described by the antisymmetric Fock space are called *fermions*, and are said to satisfy *Fermi statistics*.) It is found in Nature that the statistics a particle obeys is invariably correlated with another feature of the particle, its spin. It is found, in fact, that all particles with half-integer (i.e., half-odd-integer) spin obey Fermi statistics, while particles with integer spin obey Bose statistics. How should this fact be incorporated into quantum field theory? One could, of course, merely regard the correlation between spin and statistics as an empirical fact — a fact which can be used to choose the appropriate statistics in each case. It is natural to ask, however, whether there is some deeper theoretical reason why Nature operates as She does. Certainly, no obvious internal inconsistencies arise if we insist that Klein-Gordon and Maxwell particles be fermions, while Dirac particles be bosons. It would be desirable, however, to find some very general requirement on quantum field theories which would force the experimentally observed relation between spin and statistics. There is, in fact, such a requirement: the demand that the propagators be causal. We have seen in Sect. 23 that, with the "correct" statistics, the propagators are indeed causal. In this section, we shall indicate why the propagators for "fermion Klein-Gordon", "fermion Maxwell", and "boson Dirac" particles are not causal. These results are a special case of a core general theorem. If we require that energy be positive (to fix the complex vector space structure), and that the propagator be causal, then particles with half-integer spin must be fermions and those with integer spin bosons. We shall not discuss this general theorem further.

We begin with H_{RKG}. The (one-particle) Hilbert space is the same as before, the inner product given by (371). The operator $\phi(f)$ is still defined by (372). Now, however, we suppose that the creation and annihilation operators act on antisymmetric Fock space. Then (373) still holds, but (374) must be modified as follows:

$$\{\underline{\phi}(f), \underline{\phi}(g)\} = \hbar^2\{C(\sigma(f)), A(\sigma(g))\} + \hbar^2\{A(\sigma(f)), C(\sigma(g))\}$$
$$= \hbar^2 \left(\sigma^\alpha(f)\bar{\sigma}_\alpha(g) + \sigma^\alpha\bar{\sigma}_\alpha(f)\right)\mathbb{I}$$
$$= \hbar \left(\int_{M_\mu^+} + \int_{M_\mu^-}\right)\left(f(k)\bar{g}(k) + \bar{f}(k)g(k)\right)\mathrm{d}V\,\mathbb{I} \tag{404}$$

The "propagator" for antisymmetric statistics — the last line in (404) — is simply not causal. (Proof: Choose almost any test functions f and g with relatively spacelike supports, and evaluate the integral.) That is to say, we obtain a causal propagator in the real Klein-Gordon case if and only if we use Bose statistics. Thus, if we take causality of the propagator as a fundamental assumption, we are led to assign Bose statistics to real Klein-Gordon particles.

Now consider the complex Klein-Gordon case. If we choose Fermi statistics, (376), (377), (378), and (379)) still hold. Furthermore, (380) holds if we replace the commutators by anticommutators. For (381), however, we have

$$\{\underline{\phi}(f), \underline{\phi}^*(g)\} = \hbar^2\{C(\sigma^-(f)), A(\overline{\sigma^-(g)})\} + \hbar^2\{A(\overline{\sigma^+(f)}), C(\sigma^+(g))\}$$
$$= \hbar^2 \left(\sigma^{-\alpha}(f)\bar{\sigma}_\alpha^-(g) + \sigma^{+\alpha}(g)\bar{\sigma}_\alpha^+(f)\right)\mathbb{I}$$
$$= \frac{\hbar}{2} \left(\int_{M_\mu^+} + \int_{M_\mu^-}\right)\left(f(k)\bar{g}(k) + \bar{f}(k)g(k)\right)\mathrm{d}V\,\mathbb{I} \tag{405}$$

But the last line of (405) is not causal. Hence, in order to obtain a causal propagator, complex Klein-Gordon particles must be bosons.

If we assign Fermi statistics to H_M, (382), (383), and (384) remain valid. But (385) becomes

$$\{\underline{A}(f^a), \underline{A}(g^a)\} = \hbar^2\left(\sigma^\alpha(f^a)\bar{\sigma}_\alpha(g^a) + \sigma^\alpha(g^a)\bar{\sigma}_\alpha(f^a)\right)\mathbb{I}$$
$$= -\hbar \left(\int_{M_\mu^+} + \int_{M_\mu^-}\right)\left(f^a(k)\bar{g}_a(k) + \bar{f}^a(k)g^a(k)\right)\mathrm{d}V\,\mathbb{I} \tag{406}$$

The last line is not causal. So causality of the propagator implies Bose statistics for photons.

Finally, we attempt to impose Bose statistics on Dirac particles. Eqns. (387), (388), (389), (390), (391), and (392) remain valid. Eqn. (393) remains valid if the anticommutators are replaced by commutators. But (394) becomes

$$\left[\underline{\psi}(f^A, \bar{f}_{A'}), \underline{\psi}^*(g^A, \underline{g}_{A'})\right]$$
$$= \left(-\sigma^{-\alpha}(f^A, \bar{f}_{A'})\bar{\sigma}_\alpha^-(g^A, \bar{g}_{A'}) + \sigma^{+\alpha}(g^A, \bar{g}_{A'})\bar{\sigma}_\alpha^+(f^A, \bar{f}_{A'})\right)\mathbb{I}$$
$$= -2 \left(\int_{M_\mu^+} + \int_{M_\mu^-}\right)\left[\left(f^A(k)\bar{g}^{A'}(k) - \bar{f}^{A'}(k)g^A(k)\right)k_{AA'}\right. \tag{407}$$
$$\left. + \frac{i\mu}{\sqrt{2}}\left(\bar{f}^{A'}(k)\bar{g}_{A'}(-k) + f^A(k)g_A(-k)\right)\right]\mathrm{d}V\,\mathbb{I}$$

which, again, is not causal. Causality of the propagator implies Fermi statistics for Dirac particles.

We summarize. The requirement that energies be positive fixes the complex vector space structure of the one-particle Hilbert spaces. The additional requirement that the propagators (the commutators or anticommutators of the field operators) be causal then requires that particles with integer spin be bosons and particles with half-integer spin be fermions, at least for the our cases H_{RKG}, H_{CKG}, H_M, and H_D.

25. $\star$-Algebras

We have now obtained a number of quantum field theories of relativistic, non-interacting particles. Our approach consists, basically, of the following steps:

i) form a Hilbert space of an appropriate collection of solutions of the field equations (Klein-Gordon, Maxwell, Dirac, neutrino),

ii) introduce the corresponding symmetric or antisymmetric Fock space, and

iii) replace the original fields by operators on Fock space.

However, there exists an alternative approach, in which one begins with the field operators and their commutators (or anticommutators) as the basic objects, deriving from these the Fock space and finally the one-particle Hilbert space. While the two approaches are completely equivalent logically, they differ considerably in attitude. In particular, the alternative approach emphasizes the analogy between second quantization (the ultimate passage from fields to field operators) and first quantization (e.g., the passage from Newtonian mechanics to the Schrödinger equation). One thinks of the fields (Klein-Gordon, Maxwell, Dirac, etc.) as "classical quantities" (analogous to x and p in Newtonian mechanics) which, in the quantized version of the theory, are to become operators on some Hilbert space. This alternative approach is the one conventionally followed in textbooks. We discuss it in this section.

It is convenient to first introduce a mathematical object. A $\star$-*algebra* consists, first of all, of an associative algebra $\mathscr{A}$ (over the complexes) with unit $\mathbb{I}$. That is to say, $\mathscr{A}$ is a complex vector space on which there is defined a product, AB, between elements A and B of $\mathscr{A}$, where this product is linear in the factors and satisfies (254). Furthermore, there is an element $\mathbb{I}$ of $\mathscr{A}$ such that

$$\mathbb{I}A = A\mathbb{I} = A \tag{408}$$

for every $A \in \mathscr{A}$. (Clearly, this $\mathbb{I}$ is unique.) Furthermore, we require that, as part of the structure of a $\star$-algebra, there be given a mapping from $\mathscr{A}$ to $\mathscr{A}$ (the image of $A \in \mathscr{A}$ under this mapping written $A^\star$) subject to:

A1. For each $A \in \mathscr{A}$, $(A^\star)^\star = A$.

A2. For each $A, B \in \mathscr{A}$, $(AB)^\star = B^\star A^\star$.

A3. For each $A, B \in \mathscr{A}$, $\alpha \in \mathbb{C}$, $(\alpha A + B)^\star = \bar{\alpha}A^\star + B^\star$.

109

The standard example of a ⋆-algebra is the collection of all bounded operators which are defined everywhere on a fixed Hilbert space H. Addition is defined by adding the operators, and scalar multiplication by multiplying the operator by the complex number in the usual way. The product of two operators is defined by applying them in succession. The unit $\mathbb{I}$ is, of course, the identity operator. Finally, "⋆" represents the operation of taking the adjoint of an operator. (In fact, every bounded operator defined everywhere on a Hilbert space H has a unique adjoint. We omit the (moderately difficult) proof.) Note that it is well-defined to speak of projection, Hermitian, and unitary elements of a ⋆-algebra, for these notions involve only the structure incorporated in to the ⋆-algebra. Intuitively, one can think of a ⋆-algebra as representing "operators on a Hilbert space, but without the Hilbert space itself."

The essential idea of the approach to be described below is identical for all the relativistic field equations. It will suffice, therefore, to treat one case in detail. We select the complex Klein-Gordon fields.

The idea is to first introduce a certain ⋆-algebra $\mathscr{A}$. We suppose that, with each real test function f on Minkowski space, there is associated a pair of elements of $\mathscr{A}$, $\underline{\phi}(f)$ and $\underline{\phi}\star(f)$, which are related by the ⋆-operation. We suppose, furthermore, that $\mathscr{A}$ is generated by $\mathbb{I}$, $\underline{\phi}(f)$, and $\underline{\phi}\star(f)$ (as f runs over all test functions). That is to say, the most general element of $\mathscr{A}$ consists of a finite linear combination, with complex coefficients, of $\mathbb{I}$ and products of the $\underline{\phi}(f)$'s and $\underline{\phi}\star(f)$'s, e.g.,

$$\alpha\mathbb{I} + \beta\underline{\phi}^{\star}(f) + \gamma\underline{\phi}(g)\underline{\phi}^{\star}(k) + \delta\underline{\phi}(m)\underline{\phi}^{\star}(n)\underline{\phi}(p) \tag{409}$$

Clearly, we can take the sum or product of objects of the form (409) multiply such an object by a complex number, and take the ⋆ of such an object. Unfortunately, we still do not have quite the ⋆-algebra we require. We wish to require, in addition, that certain expressions of the form (409), while formally distinct, are to be regarded as equal as elements of $\mathscr{A}$. That is to say, we wish to impose certain relations among the elements (409) of $\mathscr{A}$. (This construction is analogous to that in which one obtains a group by postulating the existence of certain elements subject to relations. If we wished to be more formal, we would introduce an equivalence relation.) We impose the following relations:

$$\underline{\phi}(af + g) = a\underline{\phi}(f) + \underline{\phi}(g) \qquad \underline{\phi}^{\star}(af + g) = a\underline{\phi}^{\star} + \underline{\phi}^{\star}(g) \tag{410}$$

$$\underline{\phi}\left((\Box + \mu^2)f\right) = \underline{\phi}^{\star}\left((\Box + \mu^2)f\right) = 0 \tag{411}$$

$$[\underline{\phi}(f), \underline{\phi}(g)] = [\underline{\phi}^{\star}(f), \underline{\phi}^{\star}(g)] = 0 \tag{412}$$

$$[\underline{\phi}(f), \underline{\phi}^{\star}(g)] = \frac{\hbar}{2i}D(f, g)\,\mathbb{I} \tag{413}$$

where f and g are any test functions, a is any real number, and $D(f, g)$ is the Feynman propagator, (375). This completes the specification of the ⋆-algebra $\mathscr{A}$. (Although, of course, this ⋆-algebra $\mathscr{A}$ looks familiar, it is to be regarded, for the present as merely the mathematical object which results from the construction above.)

It is useful conceptually to restate the construction above from a more physical point of view. We have taken the classical field $\phi(x)$ and its complex-conjugate field $\bar{\phi}(x)$, and replaced them by operators, $\underline{\phi}(f)$ and $\underline{\phi}^\star(f)$. ("Operators" which, as yet, act on no particular Hilbert space: therefore, elements of a $\star$-algebra.) We impose on these operators a number of more or less natural conditions. We require that the operators be linear in the test functions, (410). We require that the operators, in their position dependence, satisfy the same equations as the fields they replaced, (411). We require that the $\underline{\phi}$'s commute with each other, and that the $\underline{\phi}^\star$'s commute with each other, (412). (This condition is analogous to the statement in Schrödinger theory that position operators commute with each other and momentum operators commute with each other.) We must be careful, however, that we do not have all the operators commute with each other, for the passage from classical to quantum theory involves replacing classical variables by operators with certain, nontrivial, commutation relations. What should we choose for the commutator of $\underline{\phi}(f)$ and $\underline{\phi}(g)$? We require, firstly, that the commutator be a multiple of the identity (just as the commutator of the operators x and p in the Schrödinger theory is a multiple of the identity). We require, furthermore, that $[\underline{\phi}(f), \underline{\phi}^\star(g)]$ vanish when f and g have relatively spacelike supports. (This assumption is perhaps not too unreasonable physically. When f and g have relatively spacelike supports, then, by causality, measurements made in the support of $\mathbb{I}$ should in no way affect measurements made in the support of g. Since the field operators are, in some sense, to be associated with measurements, we might expect commutativity in this case.) But these conditions ((410), (411), (412), and the assumption that the right side of (413) be a multiple of the identity which is a causal function) imply that the right side of (413) be precisely the Feynman propagator, up to an overall factor. (This statement is not difficult to prove.) Note that the "passage to a quantum theory" arises because of the assumed noncommutativity, (413).

To summarize, if we take the Klein-Gordon equation as a "classical" equation, and attempt to "quantize" it, more or less in the standard way, we are led to introduce the $\star$-algebra $\mathscr{A}$.

We next construct an inner-product space. While the construction could in principle be carried out directly in terms of the $\star$-algebra $\mathscr{A}$, it is more convenient to first introduce a second $\star$-algebra $\mathscr{B}$. This $\mathscr{B}$ is to be generated by elements $A^+(f)$, $C^+(f)$, $A^-(f)$, and $C^-(f)$, for each real test function f, subject to the following relations:

$$(A^+(f))^\star = C^+(f) \qquad (A^-(f))^\star = C^-(f) \tag{414}$$

$$\begin{aligned} A^\pm(af + g) &= aA^\pm(f) + A^\pm(g) \\ C^\pm(af + g) &= aC^\pm(f) + C^\pm(g) \end{aligned} \tag{415}$$

$$A^\pm\left((\Box + \mu^2)f\right) = C^\pm\left((\Box + \mu^2)f\right) = 0 \tag{416}$$

$$\begin{aligned} \left[C^+(f), C^+(g)\right] = \left[C^+(f), C^-(g)\right] &= \left[C^-(f), C^-(g)\right] = 0 \\ \left[A^+(f), A^+(g)\right] = \left[A^+(f), A^-(g)\right] &= \left[A^-(f), A^-(g)\right] = 0 \\ \left[A^+(f), C^-(g)\right] = \left[A^-(f), C^+(g)\right] &= 0 \end{aligned} \tag{417}$$

$$\left[A^+(f), C^+(g)\right] = \frac{\hbar}{2i} D^+(f,g)\mathbb{I} \qquad \left[A^-(f), C^-(g)\right] = \frac{\hbar}{2i} D^-(f,g)\mathbb{I} \qquad (418)$$

First note that we may regard $\mathscr{A}$ as a ⋆-subalgebra of $\mathscr{B}$. Specifically, we set

$$\begin{aligned} \underline{\phi}(f) &= \hbar\left[C^-(f) + A^+(f)\right] \\ \underline{\phi}^\star(f) &= \hbar\left[C^+(f) + A^-(f)\right] \end{aligned} \qquad (419)$$

whence each element of $\mathscr{A}$ defines an element of $\mathscr{B}$. Note that the identifications (419) indeed establish $\mathscr{A}$ as ⋆-subalgebra of $\mathscr{B}$, for (414)–(418) imply (410)–(413). In fact, although $\mathscr{B}$ is larger than $\mathscr{A}$, there is a sense in which $\mathscr{B}$ "does not add anything new" to $\mathscr{A}$. Specifically, each element of $\mathscr{B}$ can be considered as limiting case of elements of $\mathscr{A}$: $A^+(f)$, for example, is the "positive-frequency part" of $\underline{\phi}(f)$. (Just as a complex-valued solution $\phi(x)$ of the Klein-Gordon equation can be decomposed into its positive-frequency and negative-frequency parts, so can an operator-valued solution, $\phi(f)$, of the Klein-Gordon equation. It in perhaps not surprising that if one introduces enough machinery, it becomes possible to describe such a decomposition directly in terms of the ⋆-algebra.)

Why do we introduce two distinct ⋆-algebras when they carry essentially the same information? Because $\mathscr{A}$ is easier to motivate while $\mathscr{B}$ is easier to manipulate.

We now construct our inner-product space, $K_\mathscr{B}$. We postulate, first of all, the existence of an element σ_0 (the "vacuum", more commonly written $|0\rangle$). The most general element of $K_\mathscr{B}$ is to consist of the juxtaposition of an element of $\mathscr{B}$ and σ_0. We add such elements of $K_\mathscr{B}$, and multiply by complex numbers, by performing the indicated operation on $\mathscr{B}$, e.g.,

$$\alpha(A\sigma_0) + (B\sigma_0) = (\alpha A + B)\sigma_0 \qquad (420)$$

where $A, B \in \mathscr{B}$, $\alpha \in \mathbb{C}$. We wish, however, to impose on these elements a further relation, namely

$$A^+(f)\sigma_0 = 0 \qquad A^-(f)\sigma_0 = 0 \qquad (421)$$

("annihilation on the vacuum gives zero"). We now have a complex vector space. To obtain an inner-product space, we must introduce a norm. To evaluate the norm of an element of $K_\mathscr{B}$, one first formally takes the "inner product" of the element with itself. For $\alpha A^+(f)C^-(g)C^+(h)\sigma_0$, for example, one would write

$$\left(\alpha A^+(f)C^-(g)C^+(h)\sigma_0, \alpha A^+(f)C^-(g)C^+(h)\sigma_0\right) \qquad (422)$$

We now set down certain rules for manipulating such expressions. Firstly, an element of $\mathscr{B}$ which appears first on either side of the "formal inner product" can be transferred to the other side (where it must also appear first) provided that, simultaneously, it is replaced by its starred version. (That is, we mimic the usual rule for transferring an operator to the other side of an inner product.) For example, (423) can be rewritten

$$\left(\alpha C^-(g)C^+(h)\sigma_0, \alpha C^+(f)A^+(f)C^-(g)C^+(h)\sigma_0\right) \qquad (423)$$

or

$$(\alpha\bar{\alpha}A^-(g)C^+(f)A^+(f)C^-(g)C^+(h)\sigma_0, C^+(h)\sigma_0) \tag{424}$$

Secondly, one can use the commutation relations (417) and (418). For example, (423) can be rewritten

$$(\alpha C^-(g)C^+(h)\sigma_0, \alpha A^+(f)C^+(f)C^-(g)C^+(h)\sigma_0)$$
$$+ (\alpha C^-(g)C^+(h)\sigma_0, \alpha\frac{\hbar}{2i}D^+(f,f)C^-(g)C^+(h)\sigma_0) \tag{425}$$

Thirdly, one can use (421). Finally, we postulate

$$(\sigma_0, \sigma_0) = 1 \tag{426}$$

("the vacuum is normalized to unity"). By using these rules, every "norm" can be reduced to some number. This is done, roughly speaking, as follows. First use the commutators to "push the annihilation operators to the right" until they stand next to σ_0 and hence give zero. There then remain only creation operators. Each of these, in turn, is transferred to the other side of the inner product, thus becoming an annihilation operator. Each annihilation operator obtained in this way is then "pushed to the right" again, where it eventually meets σ_0 and gives zero. In this way, all the operators are eventually eliminated, leaving only the functions which appear in the commutators. Now use (426). As a simple example, we evaluate the norm of $C^+(f)\sigma_0$:

$$
\begin{aligned}
(C^+(f)\sigma_0, C^+(f)\sigma_0) &= (A^+(f)C^+(f)\sigma_0, \sigma_0) \\
&= (C^+(f)A^+(f)\sigma_0, \sigma_0) + (\frac{\hbar}{2i}D^+(f,f)\sigma_0, \sigma_0) \\
&= 0 + \frac{\hbar}{2i}D^+(f,f)(\sigma_0, \sigma_0) \\
&= \frac{\hbar}{2i}D^+(f,f)
\end{aligned} \tag{427}
$$

Thus, $K_{\mathscr{B}}$ has the structure of an inner-product space. (We shall establish shortly that the norm, defined above, is indeed positive.) In particular, if we consider only the elements of $K_{\mathscr{B}}$ which can be obtained by applying elements of $\mathscr{A}$ to σ_0 we obtain an inner-product subspace, $K_{\mathscr{A}}$, of $K_{\mathscr{B}}$. Finally, we take the completion, $\overline{K_{\mathscr{A}}}$, of $K_{\mathscr{A}}$ to obtain a Hilbert space. (In fact, $K_{\mathscr{A}}$ is dense in $K_{\mathscr{B}}$, so $\overline{K_{\mathscr{A}}} = \overline{K_{\mathscr{B}}}$.)

All these formal rules and relations sound rather mysterious. It is easy, however, to see what the resulting Hilbert space is. Consider the symmetric Fock space based on H_{CKG}. As we have mentioned, the $\star$-algebras $\mathscr{A}$ and $\mathscr{B}$ can be represented as operators on this Hilbert space. Consider now the element $(1, 0, 0, \ldots)$ (see (98)) of Fock space. It satisfies (421) and (426). Clearly, the inner-product space $K_{\mathscr{A}}$ (resp. $K_{\mathscr{B}}$) is identical to the inner-product space consisting of all elements of Fock space which can be obtained by applying elements of $\mathscr{A}$ (resp. $\mathscr{B}$) to $(1, 0, 0, \ldots)$. Thus, $K_{\mathscr{A}}$ and $K_{\mathscr{B}}$ can be considered as subspaces of our Fock space. But in fact, both these subspaces are dense in

Fock space. Hence, $\overrightarrow{K_{\mathscr{A}}}$ and $\overrightarrow{K_{\mathscr{B}}}$ are identical with symmetric Fock space based on H_{CKG}. In other words, we have simply re-obtained Fock space by a different route.

We summarize the situation. In the conventional approach, one begins with the classical fields, which are replaced by "operators" (elements of a ⋆-algebra), subject to certain commutation relations. One then assumes the existence of a vacuum, and builds an inner-product space by applying the elements of our ⋆-algebra to the vacuum. The norm is defined by formal manipulative rules, using the postulated commutators. Finally, one completes this inner-product space to obtain a Hilbert space. One has the feeling that one is "quantizing" a classical theory. We have proceeded in a rather different direction. We "looked ahead" to see what the resulting Hilbert space would be, and simply wrote it out. It turned out to be what we called the symmetric Fock based on the Hilbert space H_{CKG}, which, in turn, was based on the solutions of the original equation. We then simply defined the action of creation operators, annihilation operators, and field operators on this explicit Fock space. The resulting mathematical structures are identical — the methods of deriving this structure quite different. We have sacrificed much of the motivation to gain a certain explicitness.

26. Scattering: The S-Matrix

Our discussion so far has been restricted to free particles. That is to say, we have been dealing with systems consisting of any number of identical particles which interact neither with themselves nor with particles of any other type. While such systems provide a convenient starting point for quantum field theory, they are by themselves of very little physical interest. Particles in the real world do interact with other particles: electrons have charge, and so interact with photons; nucleons interact, at least, with π-mesons to form nuclei, etc. Furthermore, even for systems in which interactions play a minor role, the experiments which provide information about such systems must come from the interaction with other systems (i.e., the experimental apparatus). One of the most important situations — from both the theoretical and experimental points of view — in which interactions play a significant role is that of scattering. In this section, we shall set up the general framework for the description of scattering experiments.

We first recall a general principle of quantum theory. Let S_1 and S_2 represent two quantum systems which, we assume, in no way interact with each other. Let H_1 be the Hilbert space which encompasses the possible states of S_1, and H_2 the Hilbert space for S_2. It is because the systems do not interact that each is characterized by its own Hilbert space. Now suppose we introduce a new quantum system, S, which consists of S_1 and S_2 together. Note that we are not here turning on any interactions — we have merely decided to consider two systems as a single system. What is the Hilbert space of states of the system S? It is $H_1 \otimes H_2$. (Note: the tensor product, not the direct sum.) That is to say, a state of S can be obtained by taking a formal sum of formal products of states of S_1 and S_2. (Simple example: if H_1 and H_2 were both one-dimensional, so S_1 and S_2 each had essentially one state, then S should also have essentially one state. But in this example $H_1 \oplus H_2$ is two-dimensional, whereas $H_1 \otimes H_2$ is one-dimensional.) Note, incidentally, that any operator on H_1 (i.e., which acts on S_1) extends naturally to an operator on $H_1 \otimes H_2$ (i.e., extends to an operator which acts on S).

Now suppose we wish to consider a situation in which only certain types of particles will be permitted to interact — say, electrons-positrons, photons, and neutral π-mesons. We begin by writing down the Hilbert space $\mathscr{H}$ which encompasses the states of such a system when the interactions are "turned off". That is to say, we imagine a system in which our various particles co-exist but do not interact, and describe its states by $\mathscr{H}$. In our example, $\mathscr{H}$ would be the tensor product of the antisymmetric Fock space based on H_D, the symmetric Fock space based on H_M, and the symmetric Fock space based on H_{RKG}. Note that this is a purely mathematical construct. In the real world such particles would interact: we do not have the option of turning off interactions to suit our convenience.

It is in terms of this $\mathscr{H}$ that scattering processes are described. We consider the following situation. In the distant past, our particles are represented by broad and widely separated wave packets. These particles then enter a region in which the amplitudes are large and the wave packets overlap significantly. They interact. Finally, we suppose that, in the distant future, all the particles are again widely separated. We wish to describe such a process as follows. It is perhaps not unreasonable to suppose that, as one goes further and further into the past, the interactions play a smaller and smaller role. Thus, in the limit as one goes into the past, it might be possible to describe the system by an element of our non-interacting Hilbert space $\mathscr{H}$. That is, our incoming state is to be some element of $\mathscr{H}$. Similarly, in the limit as one goes into the future, i.e., for the outgoing state, one obtains some other element of $\mathscr{H}$. It is only in these "past and future limits" that $\mathscr{H}$ provides an accurate description of the state of the system. In any actual finite region of Minkowski space — and particularly where the interactions are strong — a description in terms of $\mathscr{H}$ is impossible. We therefore simply abandon, for the time being, any attempt to describe in detail what is happening while the interaction takes place. We agree that all we care to know about the interaction is simply the relation between the incoming state and the outgoing state — both elements of $\mathscr{H}$. This relation is given by some mapping S from $\mathscr{H}$ to $\mathscr{H}$.

We illustrate this idea with a classical analogy. (Caution: This analogy can be misleading if pushed too far.) Suppose we are interested in solutions of

$$(\Box + \mu^2)\phi = \phi^3 \tag{428}$$

Let $\mathscr{L}$ denote the collection of solutions of this equation which are, in some suitable sense, well-behaved asymptotically. (Note that $\mathscr{L}$ is not a vector space. It is analogous to the states of the interacting quantum system, which do form a vector space.) Let $\mathscr{H}$ denote the collection of asymptotically well-behaved solutions of the Klein-Gordon equation. Fix a time-coordinate t in Minkowski space (i.e., $\nabla_a t$ is constant, unit, and timelike). For each value of t, we define a mapping, $\Lambda(t)$, from $\mathscr{L}$ to $\mathscr{H}$. Fix t_0. Then, if $\phi(x)$ is a solution of (428), the values of ϕ and $(\nabla^a t)\nabla_a phi$ on the 3-surface $t = t_0$ are initial data for some solution of the Klein-Gordon equation, which we write $\Lambda(t)\phi$. Clearly, this mapping $\Lambda(t)$ is invertible. We now ask whether the right side of

$$S = \lim_{\substack{t_2 \to \infty \\ t_1 \to -\infty}} \Lambda(t_2)\Lambda(t_1)^{-1} \tag{429}$$

exists and is independent of our original choice of time-coordinate. If so, we obtain a mapping S from $\mathscr{H}$ to $\mathscr{H}$. This S clearly provides a great deal of information about the structure of Eqn. (428).

We now return to quantum field theory. All the information we want about the interactions is to be contained in the mapping S, called the S-*matrix*, from the non-interacting Hilbert space to itself. One could, of course, merely determine S, as best as one can, from experiments. But this would hardly represent a physical theory. Ultimately, we shall be concerned with the problem of calculating S from specific assumptions concerning the nature of the interaction. It is of interest, however, to first ask whether there are any very general properties of S which one might expect to hold merely from its physical interpretation. In fact, there are two such properties. The first is that S is an operator on $\mathscr{H}$, i.e., S is linear. I do not know of a water-tight physical argument for this assumption. It is, however, suggested by the principle of superposability in quantum theory. Let σ_1 and σ_2 be unit, orthogonal elements of H. Then $\sigma = (\sigma_1 + \sigma_2)/\sqrt{2}$ is also a unit vector in $\mathscr{H}$. A system whose incoming state is σ has probability $1/2$ that its incoming state is σ_1, and probability $1/2$ that its incoming state is σ_2. Hence, we might expect the corresponding outgoing state, $S(\sigma)$, to be the same linear combination of $S(\sigma_1)$ and $S(\sigma_2)$, i.e., we might expect to have

$$S(\sigma) = \frac{1}{\sqrt{2}} \left(S(\sigma_1) + S(\sigma_2) \right) \tag{430}$$

These considerations strongly suggest the assumption we now make: that S is a linear operator on $\mathscr{H}$. The second property of S follows from the probabilistic interpretation of states in quantum theory. Let σ be a unit vector in $\mathscr{H}$. Then, if we write $\mathscr{H}$ as a direct sum of certain of its orthogonal subspaces, the sum of the norms of the projections of σ into these subspaces is one. This fact is interpreted as meaning that the total probability of the system's being found in one of these subspaces is one. But if this σ is our incoming state, the total probability for all possible outgoing states must also be one. Hence, we might expect to have

$$\|S\sigma\| = 1 \tag{431}$$

provided $\|\sigma\| = 1$. In other words, we expect S to be a unitary operator on $\mathscr{H}$.

To summarize, the probabilities for all possible outcomes of all possible scattering experiments (involving a certain, given list of particles) are specified completely by a unitary operator S on the non-interacting Hilbert space $\mathscr{H}$. We want to find this S.

The S-matrix approach to scattering problems involves a number of physical assumptions. Among these are the following:

1. In the limit to the distant past (and distant future), the interactions have negligible influence, so the state of the actual physical system can be associated, in these limits, with elements of the non-interacting Hilbert space, $\mathscr{H}$.

2. The interaction is completely described by the S-matrix (e.g., there are no bound states.)

3. One can find short list of particles such that, if only particles which appear on this list are involved in the incoming state, then all outgoing particles will also appear on the list.

In fact, all of these assumptions are believed to be false:

1. Even in the distant past and future, particles carry a "cloud of virtual particles" which affect, for example, the observed mass. Thus, the interactions are important even in the limits. It appears, however, that these effects can be accounted for by suitably modifying the parameters (e.g., mass) which appear in the non-interacting Hilbert space $\mathcal{H}$. This procedure is associated with what is called renormalization.

2. There exist bound states, e.g., the hydrogen atom.

3. Suppose we decide that we shall allow incoming states containing only photons and electron-positrons. Then, if the energies are sufficiently large, the outgoing states could certainly include π-mesons, proton-antiproton pairs, etc. The problem is that we do not have an exhaustive list of all "elementary" particles, and so we cannot write down the "final" $\mathcal{H}$. We are forced to proceed by a series of approximations. In certain situations, interactions which produce elementary particles not included in our $\mathcal{H}$ will not play an important role. Thus, we can closely describe the physical situation by using only one or two of the many interactions between elementary particles. Whenever we write down an $\mathcal{H}$ and an S, we have a physical theory with only a limited range of validity. The larger our $\mathcal{H}$, and the more interactions included, the larger is the domain of validity of our theory.

Despite these objections, we shall, as our starting point for the discussion of interactions, use the S-matrix approach.

27. The Hilbert Space of Interacting States

We have seen in Sect. 26 that scattering phenomena are completely described by a certain unitary operator S on a Hilbert space $\mathscr{H}$. We also remarked that, since $\mathscr{H}$ represents noninteracting states, and since the states of the actual physical system are influenced by interactions, we cannot interpret $\mathscr{H}$ as encompassing the states of our system. Is it possible, then, to construct a Hilbert space $\mathscr{L}$ which does represent the states of the interacting system? The answer is yes (at least, for scattering states), provided we accept a sufficiently loose interpretation for the word "construct."

What features would we expect for a Hilbert snare which is to represent the "interacting states of the system"? Firstly, comparing $\mathscr{L}$ and $\mathscr{H}$ in the distant past, we might expect to have an isomorphism $\Lambda_{\text{in}} : \mathscr{L} \to \mathscr{H}$ between $\mathscr{L}$ to $\mathscr{H}$. (See the example on p. 116.) (An *isomorphism* between two Hilbert spaces is a mapping from one to the other which is one-to-one, onto, linear, and norm-preserving. Clearly, any isomorphism has an inverse, which is itself an isomorphism.) Similarly, we would expect to have a second isomorphism $\Lambda_{\text{out}} : \mathscr{L} \to \mathscr{H}$. Finally, from the definition of the S-matrix, we would expect to have

$$S = \Lambda_{\text{out}} \Lambda_{\text{in}}^{-1} \tag{432}$$

Fix $\mathscr{H}$ and S. A triple, $(\mathscr{L}, \Lambda_{\text{in}}, \Lambda_{\text{out}})$ consisting of a Hilbert space $\mathscr{L}$ and two isomorphisms, Λ_{in} and Λ_{out}, from $\mathscr{L}$ to $\mathscr{H}$, subject to (432), will be called on *interaction space*. How many essentially different interaction spaces are there for a given $\mathscr{H}$, S? In fact, there is just one, in the following sense: Let $(\mathscr{L}', \Lambda'_{\text{in}}, \Lambda'_{\text{out}})$ and $(\mathscr{L}, \Lambda_{\text{in}}, \Lambda_{\text{out}})$ be two interaction spaces for $\mathscr{H}$, S. Then there exists a unique isomorphism Ψ from $\mathscr{L}$ to $\mathscr{L}'$ such that

$$\Lambda_{\text{in}} = \Lambda'_{\text{in}} \cdot \Psi \qquad \Lambda_{\text{out}} = \Lambda'_{\text{out}} \cdot \Psi \tag{433}$$

(Proof: Evidently, we must choose $\Psi = {\Lambda'_{\text{in}}}^{-1} \Lambda_{\text{in}}$, which, by (432), is the same as $\Psi = {\Lambda'_{\text{out}}}^{-1} \Lambda_{\text{out}}$.) That is to say, the interaction space is unique up to isomorphism (which, of course, is as unique as we could expect it to be.)

All this looks rather pedagogical. After all, $\mathscr{L}$ is just another copy of $\mathscr{H}$, so why don't we just say so instead of speaking of "triples," etc.? The point is that the interaction space is more than just another cony of $\mathscr{H}$ — it also contains,

119

as part of its structure, certain isomorphisms from $\mathscr{L}$ to $\mathscr{H}$. As a consequence, only certain portions of the (extensive) structure on $\mathscr{H}$ can be carried over, in a natural way, to $\mathscr{L}$. Examples of structure on $\mathscr{H}$ — which arise from the way in which $\mathscr{H}$ was constructed — are:

1. The total charge operator on $\mathscr{H}$.

2. The "total number of photons" operator on $\mathscr{H}$ (if, say, $\mathscr{H}$ happens to include photons).

3. The projection operator onto photon states (eliminating all other types of particles).

4. The operators which arise from the action of the restricted Poincaré group on $\mathscr{H}$.

5. The "number of baryons minus number of anti-baryons" operator.

6. The creation and annihilation operators of various particles in various states.

7. The field operators on $\mathscr{H}$.

The important point is that, in every case, the "additional structure" on $\mathscr{H}$ can be described by giving an operator on $\mathscr{H}$. The question of transferring structure from $\mathscr{H}$ to $\mathscr{L}$ reduces, therefore, to the following: under what conditions does an operator A on $\mathscr{H}$ define, in a natural way, an operator on $\mathscr{L}$? In fact, given an operator A on $\mathscr{H}$, there are two natural operators on $\mathscr{L}$, namely,

$$\Lambda_{\text{in}}^{-1} A \Lambda_{\text{in}} \qquad \Lambda_{\text{out}}^{-1} A \Lambda_{\text{out}} \tag{434}$$

In other words, we can carry A from $\mathscr{H}$ to $\mathscr{L}$ via either of the isomorphisms Λ_{in} or Λ_{out}. Which of (434) should we choose? There would he no choice if these two operators were equal. Thus, an operator A on $\mathscr{H}$ leads to a unique operator on the interaction space $\mathscr{L}$ provided A is such that the two operators (434) are equal. Using (432), this is equivalent to the condition

$$[S, A] = 0 \tag{435}$$

This, of course, is the result we expect. It is only properties of $\mathscr{H}$ which are invariant under the interaction (i.e., operators which commute with the S-matrix) which lead unambiguously to properties of the interaction states.

To summarize, structure on the interaction space is obtained from operators on $\mathscr{H}$ which commute with the S-matrix.

The operators which commute with S characterize what are called *conserved quantities*. They include charge, baryon number, lepton number, momentum, angular momentum, etc. Operators such as 2, 3, 6, and 7 above will not commute with S. They describe quantities which are not conserved in interactions.

28. Calculating the S-Matrix: An Example

We shall soon begin writing down formulae for the S-matrix. Unfortunately, these formula are rather complicated. They contain large numbers of terms, sums and integrals whose convergence is doubtful, and symbols whose precise meaning is rather obscure. We wish to avoid encountering all of these problems simultaneously. It is convenient, therefore, to first study a simpler example — a problem in which some of the features of the S-matrix formulae are exhibited, and in which some, but only some, of the difficulties are seen. We discuss such an example in the present section.

Let H he a fixed Hilbert space. Let $K(t)$ be a one-parameter family of bounded operators defined everywhere on H. That is, for each real number t, $K(t)$ is an operator on H. Suppose furthermore that $K(t) = 0$ unless t is in some finite interval. That is, suppose that there are numbers $t_i < t_f$ such that $K(t) = 0$ for $t \geq t_f$ and $t \leq t_i$. We are interested in studying curves in H, i.e., one-parameter families $\sigma(t)$ of elements of H, which satisfy the equation

$$-\frac{\hbar}{i}\frac{\mathrm{d}}{\mathrm{d}t}\sigma(t) = K(t)\sigma(t) \tag{436}$$

(Note: Derivatives and integrals of one-parameter families of elements of a Hilbert space, and operators on a Hilbert space, are defined by the usual limiting procedure.) Let $\sigma(t)$ satisfy (436). Then, since $K(t) = 0$ for $t \leq t_i$, $\sigma(t)$ is a constant element of H, σ_i, for $t \leq t_i$. Similarly, $\sigma(t) = \sigma_f$ for $t \geq t_f$. Clearly, a solution of (436) is completely and uniquely determined by σ_i, and σ_f is a linear function of σ_i. We write

$$\sigma_f = S\sigma_i \tag{437}$$

where S is some operator on H. The problem is to find an expression for S in terms of $K(t)$.

We first consider a special case in which the solution is easy. Suppose that all the $K(t)$'s commute with each other, i.e., $[K(t), K(t')] = 0$ for any t and t'. Then

$$\sigma(t) = \left[\exp\left(-\frac{i}{\hbar}\int_{t_i}^{t}\mathrm{d}\tau K(\tau)\right)\right]\sigma_i \tag{438}$$

is clearly a solution of (436). (Note: If $A(t)$ is a one-parameter (differentiable) family of operators on H, then $\frac{\mathrm{d}}{\mathrm{d}t}\exp A(t) = (\exp A(t))\frac{\mathrm{d}}{\mathrm{d}t}A(t)$ only when $A(t)$

and $\frac{d}{dt}A(t)$ commute.) Therefore,

$$
\begin{aligned}
S &= \exp\left(-\frac{i}{\hbar}\int_{t_i}^{t_f} d\tau K(\tau)\right) \\
&= \mathbb{I} + \left(-\frac{i}{\hbar}\int_{t_i}^{t_f} d\tau K(\tau)\right) + \frac{1}{2!}\left(-\frac{i}{\hbar}\int_{t_i}^{t_f} d\tau K(\tau)\right)^2 \\
&\qquad\qquad + \frac{1}{3!}\left(-\frac{i}{\hbar}\int_{t_i}^{t_f} d\tau K(\tau)\right)^3 + \cdots \quad (439)
\end{aligned}
$$

(The exponential of an operator is, of course, defined by the second equality in (439). We ignore, for the time being, questions of the convergence of such series.) Thus, when the $K(t)$'s commute, S is given by the relatively simple expression (439).

Now suppose that the $K(t)$'s do not commute. Integrating (436), we rewrite it as an integral equation:

$$
\sigma(t) = \sigma_i - \frac{i}{\hbar}\int_{t_i}^{t} d\tau K(\tau)\sigma(\tau) \tag{440}
$$

We shall solve (440), at least formally, using a sequence of approximations. We begin with a trial solution, $\sigma_0(t)$. We substitute this $\sigma_0(t)$ into the right side of (440), and denote the result by $\sigma_1(t)$. We now take $\sigma_1(t)$ as our next trial solution. Substituting it into the right side of (440) to obtain $\sigma_2(t)$, etc. Thus, the general formula for passing from one trial solution to the next is

$$
\sigma_{n+1}(t) = \sigma_i - \frac{i}{\hbar}\int_{t_i}^{t} d\tau K(\tau)\sigma_n(\tau) \tag{441}
$$

As our initial trial solution, we take $\sigma_0(t) = \sigma_i$, a constant. The hope is that the resulting $\sigma_n(t)$ will, as $n \to \infty$, converge, in a suitable sense, to a solution $\sigma(t)$ of (436). Using (441) successively, we have

$$
\begin{aligned}
\sigma_1(t) &= \sigma_i - \frac{i}{\hbar}\int_{t_i}^{t} d\tau_1 K(\tau_1)\sigma_i \\
\sigma_2(t) &= \sigma_i - \frac{i}{\hbar}\int_{t_i}^{t} d\tau_1 K(\tau_1)\sigma_i + \left(-\frac{i}{\hbar}\right)^2\int_{t_i}^{t} d\tau_1 \int_{t_i}^{\tau_1} d\tau_2 K(\tau_1)K(\tau_2)\sigma_i
\end{aligned}
\tag{442}
$$

or, more generally,

$$
\sigma_n(t) = \left[\sum_{m=0}^{n}\left(-\frac{i}{\hbar}\right)^m \int_{t_i}^{t} d\tau_1 \int_{t_i}^{\tau_1} d\tau_2 \cdots \int_{t_i}^{\tau_{m-1}} d\tau_m K(\tau_1)\cdots K(\tau_m)\right]\sigma_i \tag{443}
$$

Thus, our formal limiting solution is

$$
\sigma_n(t) = \left[\sum_{m=0}^{\infty}\left(-\frac{i}{\hbar}\right)^m \int_{t_i}^{t} d\tau_1 \int_{t_i}^{\tau_1} d\tau_2 \cdots \int_{t_i}^{\tau_{m-1}} d\tau_m K(\tau_1)\cdots K(\tau_m)\right]\sigma_i \tag{444}
$$

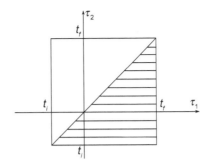

Figure 28.1: Region of integration in (447).

Indeed, if we substitute (444) into (436), and ignore questions of convergence of sums and the validity of interchanging the order of differentiation and summation, we obtain an identity. Thus, our formal expression for S is

$$S = \sum_{m=0}^{\infty} \left(-\frac{i}{\hbar}\right)^m \int_{t_i}^{t} d\tau_1 \int_{t_i}^{\tau_1} d\tau_2 \cdots \int_{t_i}^{\tau_{m-1}} d\tau_m K(\tau_1) \cdots K(\tau_m) \qquad (445)$$

It is convenient to recast (445) into a form which more closely resembles (439). The idea is to eliminate integrals whose limits of integration lie between t_i and t_f, i.e., to have all integrals be over the full range from t_i to t_f. Explicitly, the first few terms of (445) are

$$S = \mathbb{I} + \left(-\frac{i}{\hbar}\right) \int_{t_i}^{t_f} d\tau_1 K(\tau_1) + \left(-\frac{i}{\hbar}\right)^2 \int_{t_i}^{t_f} d\tau_1 \int_{\tau_1}^{t_f} d\tau_2 K(\tau_1)K(\tau_2)$$

$$+ \left(-\frac{i}{\hbar}\right)^3 \int_{t_i}^{t_f} d\tau_1 \int_{\tau_1}^{t_f} d\tau_2 \int_{\tau_2}^{t_f} d\tau_3 K(\tau_1)K(\tau_2)K(\tau_3) + \cdots \qquad (446)$$

The first two terms on the right in (446) are already in the desired form. However, the third term on the right,

$$\left(-\frac{i}{\hbar}\right)^2 \int_{t_i}^{t_f} d\tau_1 \int_{\tau_1}^{t_f} d\tau_2 K(\tau_1)K(\tau_2) \qquad (447)$$

is not. The region of integration in (447) is shown in the figure. The idea is to reverse the orders of the two integrations, while keeping the actual region over which the integration is performed — the shaded region in Figure 28.1 — unchanged. Thus, the expression (447) is equal to

$$\left(-\frac{i}{\hbar}\right)^2 \int_{t_i}^{t_f} d\tau_2 \int_{\tau_2}^{t_f} d\tau_1 K(\tau_1)K(\tau_2) \qquad (448)$$

We next reverse the roles of the integration variables, τ_1 and τ_2, in (448) to obtain

$$\left(-\frac{i}{\hbar}\right)^2 \int_{t_i}^{t_f} d\tau_1 \int_{\tau_1}^{t_f} d\tau_2 K(\tau_2)K(\tau_1) \qquad (449)$$

Finally, adding (447) and (449), we find that (447) is equal to

$$\frac{1}{2}\left(-\frac{i}{\hbar}\right)^2 \int_{t_i}^{t_f} d\tau_1 \int_{t_i}^{t_f} d\tau_2 T\left[K(\tau_1), K(\tau_2)\right] \tag{450}$$

where we have defined

$$T\left[K(\tau_1), K(\tau_2)\right] = \begin{cases} K(\tau_1)K(\tau_2) & \text{if } \tau_1 \geq \tau_2 \\ K(\tau_2)K(\tau_1) & \text{if } \tau_2 \geq \tau_1 \end{cases} \tag{451}$$

A similar procedure can be applied to each successive term in (446). The nth term is equal to

$$\frac{1}{n!}\left(-\frac{i}{\hbar}\right)^n \int_{t_i}^{t_f} d\tau_1 \cdots \int_{t_i}^{t_f} d\tau_n T\left[K(\tau_1), \ldots, K(\tau_n)\right] \tag{452}$$

where $T[K(\tau_1), K(\tau_2), \ldots, K(\tau_n)]$ is defined to be the product of these operators, but arranged in the order in which the operator associated with the smallest τ_m is placed on the right, the operator associated with the next-smallest τ_m is placed next, etc. This $T[K(\tau_1), K(\tau_2), \ldots, K(\tau_n)]$ is called the *time-ordered product*. Thus, our final formal expression for S is

$$S = \sum_{n=0}^{\infty} \frac{1}{n!}\left(-\frac{i}{\hbar}\right)^n \int_{t_i}^{t_f} d\tau_1 \int_{t_i}^{t_f} d\tau_2 \cdots \int_{t_i}^{t_f} d\tau_n T\left[K(\tau_1), \ldots, K(\tau_n)\right] \tag{453}$$

Note that, if all the $K(\tau)$'s commute, then the time-ordering is irrelevant, and (453) reduces to (439).

To summarize, the only modification of (439) required when the operators do not commute is that the products of operators in the multiple integrals must be time-ordered.

Finally, note that if all the $K(\tau)$'s are Hermitian, then, from (436),

$$\begin{aligned} \frac{d}{dt}\left(\sigma(t), \sigma(t)\right) &= \left(\frac{d}{dt}\sigma, \sigma\right) + \left(\sigma, \frac{d}{dt}\sigma\right) \\ &= \left(-\frac{i}{\hbar}K\sigma, \sigma\right) + \left(\sigma, -\frac{i}{\hbar}K\sigma\right) \\ &= \frac{i}{\hbar}\left[(K\sigma, \sigma) - (\sigma, K\sigma)\right] = 0 \end{aligned} \tag{454}$$

Hence, S must be a unitary operator. Unitarity is obvious in (439) — rather less so in (453).

Our formulae for the S-matrix in field theory will also involve infinite series of multiple integrals of time-ordered products of Hermitian operators.

29. The Formula for the S-Matrix

In this section we shall write down the formula for the S-matrix in terms of a certain (as yet unspecified) operator field on Minkowski space. While we shall in no sense "derive" that our expression for S is correct, it will be possible, at least, to show that the formula is a reasonable guess. We rely heavily on the discussion in Sect. 28.

Consider Eqn. (453). We wish to write down an analogous formula for S in quantum field theory. Since, first of all, S is to be an operator on the non-interacting Hilbert space we must take the K's to be operators on $\mathscr{H}$. Secondly, S should be unitary: we therefore take the K's to de, Hermitian. By what should we replace the interaction variables - the τ's in Eqn. (453)? If we think of τ in Sect. 28 as representing a "time", then a natural replacement would be position x in Minkowski space-time. The integrals in (453) would then extend over all of Minkowski space. (Note: This is the reason why it was convenient, in Sect. 28, to obtain an expression in which the integrals extended over the entire t-range from t_i to t_f.) Thus, we are led to consider the interaction is described by a certain Hermitian operator field, $\underline{K}(x)$, which depends on position x in Minkowski space, and acts on $\mathscr{H}$. The S-matrix will then be given by the formula

$$S = \sum_{n=0}^{\infty} \frac{1}{n!} \left(-\frac{i}{\hbar} \right)^n \int \mathrm{d}V_1 \cdots \int \mathrm{d}V_n T\left[\underline{K}(x_1), \ldots, \underline{K}(x_n)\right] \qquad (455)$$

where $x_1, x_2, \ldots$ represent points in Minkowski space, $\mathrm{d}V_1, \mathrm{d}V_2, \ldots$ the corresponding volume elements in Minkowski space, and all integrals are over all of Minkowski space. Note that Eqn. (453) already suffers from one difficulty: the question of the convergence of the infinite series. In writing (455) we have retained that difficulty, and, in fact , introduced a second one: the question of the convergence of the integrals. (The integrals in (453) are all over compact regions.)

In fact, there is a second problem with (455) which was not encountered in (453), namely, the question of what the time-ordering operator T is to mean in (455). In (453), T means that the $K(\tau)$ operators are to be placed in the order of decreasing τ-values. Unfortunately, in the passage from a "time" τ to position x in Minkowski space-time, the natural ordering is destroyed. There is, however, one case in which points in Minkowski space can be ordered: we agree that x_2 exceeds x_1, $x_2 > x_1$, if $x_2 - x_1$ (i.e., the position vector of x_2 relative to x_1) is future-directed and timelike or null. Hence, for the region of integration in (455) for which all the $x_1, \ldots, x_n$ can be ordered in this way, T has a well-defined meaning, and hence the integral makes sense. (More explicitly, the τ's are totally ordered, while points in (time-oriented) Minkowski space are only partially ordered.) Clearly, there is no Poincaré-invariant way to "time-order" $\underline{K}(x_1)$ and $\underline{K}(x_2)$ when $x_2 - x_1$ is spacelike. How, then, are we to give a meaning to (455)? One way of doing this would be through an additional condition on the $\underline{K}(x)$. We simply assume that the ordering of $\underline{K}(x_1)$ and $\underline{K}(x_2)$ is irrelevant when $x_2 - x_1$ is spacelike. That is to say, a natural way of forcing a meaning for (455) is to assume that the $\underline{K}(x)$ have the property

$$[\underline{K}(x), \underline{K}(x')] = 0 \qquad \text{for } x - x' \text{ spacelike} \qquad (456)$$

We include (456) as a requirement on our $\underline{K}(x)$. Thus, if $x_1, x_2, \ldots, x_n$ are points in Minkowski space, we define $T[\underline{K}(x_1), \underline{K}(x_2), \ldots, \underline{K}(x_n)]$ to be the product of these operators, placed in an order such that, if $x_i - x_j$ is timelike or null and future-directed, then $\underline{K}(x_i)$ appears before $\underline{K}(x_j)$ in the product. Clearly, there always exists at least one ordering having this property. Furthermore, (456) implies that all such orderings yield the same operator. Thus, $T[\underline{K}(x_1), \ldots, \underline{K}(x_n)]$ is a well-defined operator.

To summarize, the interaction is to be described by giving a Hermitian operator field $\underline{K}(x)$ on $\mathscr{H}$ which satisfies (456). The S-matrix is to be expressed in terms of $\underline{K}(x)$ by (455). The formal expression (455) is unsatisfactory insofar as we have investigated the convergence of neither the infinite series nor the integrals themselves.

The standard textbooks give a more detailed, but, I feel, no more satisfactory, argument for (455). One chooses a time-coordinate t in Minkowski space, and writes

$$\underline{H}(t) = \int_{t=\text{const.}} \underline{K}(x) \qquad (457)$$

for the "Hamiltonian". One then imagines a "state vector" in the Hilbert space $\mathscr{H}$, $\sigma(t)$, which depends on the time-coordinate. One writes the "Schrödinger equation",

$$-\frac{\hbar}{i}\frac{d}{dt}\sigma(t) = \underline{H}(t)\sigma(t) \qquad (458)$$

The argument of Sect. 28 then yields (455), where T refers to the ordering induced by the time-coordinate t. Finally Poincaré invariance (i.e., the condition that S be independent of our original choice of t) requires (456). This argument assumes, implicitly, that the "states during the interaction" are described by $\mathscr{H}$. We have deviated only slightly from this conventional argument. We isolated the simplest and clearest part in Sect. 28, and guessed the rest.

30. Dimensions

With each of the various fields and operators we have introduced, there is associated a corresponding physical dimension. We shall determine these dimensions in the present section.

Recall that we have set the speed of light equal to one, so length and time have the same units. (E.g., we measure distance in light-seconds.) We may therefore take as our fundamental units a mass (m) and a time (t). Then Planck's constant h has dimensions mt. We assign to position vectors in Minkowski space dimensions t, so the derivative in Minkowski space has dimensions t^{-1}, and the wave operator $\Box$ dimensions t^{-2}. (Raising, lowering, and contracting indices does not affect dimensions.) The quantity μ which appears in the Klein-Gordon and Dirac equations therefore has dimensions t^{-1}. Position vectors in momentum space have dimensions t^{-1}. Finally, the volume element on the mass shell has dimensions t^{-1} (see (14).)

The rule for determining the dimensions to be associated with a classical field is the following: consider an element of the Hilbert space which has norm unity, and work back to determine the dimensions of the corresponding field. Consider first the (real or complex) Klein-Gordon case. A unit vector in the Hilbert space is represented by a function $\phi(k)$ on M_μ which satisfies

$$\frac{1}{\hbar} \int_{M_\mu} \phi(k)\overline{\phi(k)} \, dV_\mu = 1 \tag{459}$$

Therefore, $\phi(k)$ has dimensions $m^{1/2}t^{3/2}$. But

$$\phi(x) = \int \phi(k)e^{ik_b x^b} \, dV \tag{460}$$

and so the Klein-Gordon field has dimensions $m^{1/2}t^{-1/2}$. For the Dirac case, Eqn. (388) implies that $(\xi^A(k), \eta_{A'}(k)$ has dimensions $t^{1/2}$. Then (309) and (310) imply that $(\xi^A(k), \eta_{A'}(k)$ has dimensions $t^{3/2}$. The neutrino fields have the same dimensions. Finally, for the Maxwell case, (382) implies that $A_a(k)$ has dimensions $m^{1/2}t^{3/2}$, whence, from (178), $A_a(x)$ has dimensions $m^{1/2}t^{-1/2}$. (In the Maxwell case, one has a simple independent check on the dimensions. The dimensions above for the vector potential imply that electric and magnetic fields have dimensions $m^{1/2}t^{3/2}$ which agrees, of course, with the dimensions from classical electrodynamics.)

The other quantities whose dimensions are of particular interest are the field operators. To assign dimensions to these operators, we must decide what dimensions are to be associated with the test fields. Recall that the role of the test fields is to smear out the (undefined) operators associated with points of Minkowski space, e.g.,

$$\underline{\phi}(f) = \int f(x)\underline{\phi}(x) \, dV \tag{461}$$

Thus, we can think of a test field as a "smearing density." We therefore take all test fields to have dimensions t^{-4}. (That is, we require that $\underline{\phi}(f)$ and the (undefined) $\underline{\phi}(x)$ have the same dimensions.) With this convention, the determination of the dimensions of the field operators is straightforward. Consider first the Klein-Gordon case. If $f(x)$ is a test field (dimensions t^{-4}), then, from

$$f(x) = \int f(k)e^{ik_b x^b} \, dV \tag{462}$$

$f(k)$ is dimensionless. But a dimensionless element of our Hilbert space, (459), defines a $\phi(k)$ with dimensions $m^{1/2}t^{3/2}$. Therefore, the element of our Hilbert space associated with this $f(k)$, $\sigma(f)$, has dimensions $m^{-1/2}t^{-3/2}$. Thus, the creation and annihilation operators have dimensions $m^{-1/2}t^{-3/2}$. But

$$\underline{\phi}(f) = \hbar C(\sigma(f)) + \hbar A(\sigma(f)) \tag{463}$$

(say, for real Klein-Gordon fields), and so the field operators, $\underline{\phi}(f)$, have dimensions $m^{1/2}t^{1/2}$.   In   the   Dirac   case,   the   test   fields, $(f^A(x), f_{A'}(x))$ have dimensions t^{-4}, whence $f^A(k)$ is dimensionless. The corresponding pair of functions on the mass shell, (389), therefore has dimensions t^{-1}. Hence, the corresponding elements of our Hilbert space, $\sigma(f^A, f_{A'})$, have dimensions $t^{-3/2}$. This, then, is the dimensions of the creation and annihilation operators. Finally, from (390), the field operators in the Dirac case have dimensions $t^{-3/2}$. Similarly, in the Maxwell case, $f^a(x)$ has dimensions t^{-4}, $f^a(k)$ dimensionless, $\sigma(f^a)$ dimensions $m^{-1/2}t^{-3/2}$. The creation and annihilation operators therefore have dimensions $m^{-1/2}t^{-3/2}$, and so by (383), the Maxwell field operators have dimensions $m^{1/2}t^{-1/2}$.

Note that, in every case, the classical fields and the corresponding field operators have the same dimensions.

31. Charge Reversal

An important tool, both for constructing and analyzing interactions, is the discrete symmetries — charge, parity, and time reversal. We begin with charge reversal.

Let $\mathscr{H}$ be a non-interacting Hilbert space, and suppose we are given an operator S, the S-matrix, on $\mathscr{H}$. What does it mean to say that "the interaction described by S is invariant under charge reversal?" Roughly speaking, this means that if we replace any in-state, $\sigma \in \mathscr{H}$, by "the same state, but with all particles replaced by their antiparticles," then the corresponding out-states are again "the same, but with particles replaced by antiparticles." Thus, we are led to try to give a meaning to the notion "the same state (in $\mathscr{H}$), but with particles replaced by antiparticles." Let us suppose that this mapping from non-interacting states to non-interacting states is accomplished by come operator C on $\mathscr{H}$, so that $C\sigma$ represents the same state as σ, but with particles replaced by antiparticles. Then the statement that the interaction (described by S) is invariant under charge reversal reduces to the condition

$$SC\sigma = CS\sigma \tag{464}$$

for any $\sigma \in \mathscr{H}$. In other words, invariance under charge reversal is expressed mathematically by the condition that S and C (both operators on $\mathscr{H}$) commute.

In general, there will be a number of different operators C which could be interpreted as effecting the replacement of particles by antiparticles. There is no obvious, unambiguous way of translating this physical notion into a mathematical operator. We shall therefore proceed as follows. We first write down a list of properties which reflect the intuitive idea of "replacing particles by antiparticles, but not otherwise changing the state." In general, there will he a moderately large class of C's which satisfy these criteria. Then, for each interaction, we look for an operator C which satisfies our criteria and which, in addition, commutes with the S-matrix. If such a C exists, we say that our interaction is invariant under charge-reversal. The point is that *any* operator which commutes with the S-matrix is valuable. We regard the words "charge reversal" as merely suggesting a particularly fertile area in which such operators might be found. This philosophy is important:

i) there is no natural, *a priori* charge-reversal operator;

ii) one sets up a class of possible charge-reversal operators, and then selects from this class depending on what the interaction is,

iii) if no operator in this class commutes with the S-matrix, there is little point in considering charge-reversal for that interaction.

(The third point is somewhat over-stated, and will be modified slightly later.)

The first condition on C is that it should not mix up particles of different types. That is to say, C should commute with the total number operators on each of the Fock spaces which make up $\mathcal{H}$. Therefore, C can be considered as an operator on each of the Fock spaces separately. The assumption that C commutes with the total number operators implies, furthermore, that C can be decomposed into an operator on the one-particle Hilbert space, an operator on the two-particle Hilbert space, etc. We next assume that the action of C on the many-particle states can be obtained from the action on the one-particle states as follows. Let H be one of our one-particle Hilbert spaces (e.g., H_{RKG}, H_{CKG}, H_M, H_D). Then an element of the corresponding Fock space consists of a string

$$(\xi, \xi^\alpha, \xi^{\alpha\beta}, \dots) \tag{465}$$

of tensors over H. The operator C on the one-particle Hilbert space H can be written, in the index notation, as $C^\alpha{}_\beta$: the result of applying $C^\alpha{}_\beta$ to an element ξ^β of H is written $C^\alpha{}_\beta \xi^\beta$. We assume that the action of C on the element (465) of Fock space is

$$(\xi, C^\alpha{}_\beta \xi^\beta, C^\alpha{}_\mu C^\beta{}_\nu \xi^{\mu\nu}, \dots) \tag{466}$$

This is a quite reasonable assumption: if we know what charge-reversal means on a one-particle state, we assume that, for a two-particle state, the effect of charge reversal is to "apply charge-reversal to each of the particles individually."

Thus, we are led to distinguish a class of charge-reversal operators on each of our one-particle Hilbert spaces, H_{RKG}, H_M, etc.

It is convenient to introduce some definitions. A mapping T from a Hilbert space H to itself is said to be *antilinear* if

$$T(\alpha\sigma + \tau) = \bar{\alpha}T(\sigma) + T(\tau) \tag{467}$$

for any $\sigma, \tau \in H$, $\alpha \in \mathbb{C}$. (Alternatively, T could be considered as a linear mapping from H to $\bar{H}$.) We shall sometimes refer to an antilinear mapping as an antilinear operator. The word "operator" alone means "linear operator." A linear or antilinear operator T is said to be *norm-preserving* if

$$\|T\sigma\| = \|\sigma\| \tag{468}$$

for every $\sigma \in H$. Eqn. (468) immediately implies that, for any $\sigma, \tau \in H$,

$$(T\sigma, T\tau) = (\sigma, \tau) \tag{469}$$

$$\text{or} \qquad (T\sigma, T\tau) = (\tau, \sigma) \tag{470}$$

according as T is linear or antilinear, respectively. As we have remarked, a linear, norm-preserving operator is called unitary. An antilinear, norm-preserving operator is said to be *antiunitary*.

Let H be one of the Hilbert spaces H_{RKG}, H_{CKG}, H_M, or H_D. A linear or antilinear operator C on H will be called a *charge-reversal operator* if

1. C is norm-preserving.

2. C commutes with all the unitary operators on H which arise from the action of the restricted Poincaré group on H.

3. C, applied to a positive- (resp., negative-) frequency element of H, yields a negative- (resp., positive-) frequency element.

These conditions are to reflect the intuitive that "the state is changed only in that particles are replaced by antiparticles." Conditions 1 and 3 are clearly reasonable. (The passage from positive-frequency to negative-frequency is the passage from particles to anti-particles.) Condition 2 ensures that quantities such as the locations and momenta of particles are unchanged under C. Note that, if C is a charge-reversal operator, and α is a complex number with $|\alpha| = 1$, then αC is also a charge-reversal operator. (One could, conceivably, impose the further condition $C^2 = 1$. We shall not do so.) Note that Condition 1 implies that C is also norm-preserving on the non-interacting Hilbert space $\mathscr{H}$.

Before discussing examples of charge-reversal operators, we establish the following result: C must be linear rather than antilinear. Let r^a be a constant, unit, future-directed, timelike vector field in Minkowski space. Then the energy operator associated with r^a is

$$E = \frac{i}{\hbar} r^a \underline{L}_a \tag{471}$$

where $r^a \underline{L}_a$ is the operator which comes from the unitary transformation associated with the Poincaré transformation (a translation) generated by r^a. It is essential, in (471), that the $i/\hbar$ appear explicitly, so that r^a is simply the first-order difference between a unitary operator and the identity operator. We assume that C is antiunitary, and obtain a contradiction. For each of our Hilbert spaces, the expectation value of the energy $\underline{E}$ in any state (and, in particular, in the state $C^{-1}\sigma$, for $\sigma \in H$) is non-negative:

$$(C^{-1}\sigma, \underline{E}C^{-1}\sigma) \geq 0 \tag{472}$$

But, from (470), this implies

$$(C\underline{E}C^{-1}\sigma, \sigma) \geq 0 \tag{473}$$

Write

$$C\underline{E}C^{-1} = \left(C\frac{i}{\hbar}C^{-1}\right)(Cr^a\underline{L}_aC^{-1}) \tag{474}$$

Condition 2 above implies that $C(r^a\underline{L}_a)C^{-1}$. The assumption that C is antiunitary implies $C(i/\hbar)C^{-1} = -i/\hbar$. Thus, we have

$$(C^{-1}\sigma, \underline{E}C^{-1}\sigma) = -(\underline{E}\sigma, \sigma) \tag{475}$$

But this is a contradiction, for the left side is non-negative and the right side non-positive. Therefore, C cannot be anti-linear.

The question of the uniqueness of charge-reversal operators is settled by the following fact: let O be a bounded operator defined everywhere on H (one of our four one-particle Hilbert spaces). Suppose that O commutes with the unitary operators which arise from the action of the restricted Poincaré group and, furthermore, that O takes positive- (resp., negative-) frequency to positive- (resp., negative-) frequency states. Then O is a multiple of the identity. I know of no simple proof. (The statement is essentially an infinite-dimensional generalization of Schur's Lemma (p. 73).) Now suppose that C and C' are charge-reversal operators. Then $C'C^{-1}$ satisfies the conditions above, and hence must be a multiple of the identity. It follows that $C' = \alpha C$, where α is a complex number with $|\alpha| = 1$. Thus, having found one charge-reversal operator on our Hilbert space H, we have found them all.

We begin with H_{RKG}. In this case, the only positive-frequency or negative-frequency solution is the zero solution. (That is to say, H_{RKG} describes particles which are identical with their antiparticles.) Hence, Condition 3 is empty. Therefore, the identity is a possible charge-reversal operator. We conclude that the most general charge-reversal operator on H_{RKG} is $\alpha\mathbb{I}$, with $|\alpha| = 1$.

For H_{CKG}, one charge-reversal operator is given by

$$\phi(x) \to \bar{\phi}(x) \tag{476}$$

or, in momentum space, by

$$\phi(k) \to \bar{\phi}(-k) \tag{477}$$

That this operator is unitary rather than antiunitary follows from our complex vector-space structure on H_{CKG} (see Sect. 12). Thus, the most general charge-reversal operator on H_{CKG} is given by

$$\phi(x) \to \alpha\bar{\phi}(x) \tag{478}$$

with $|\alpha| = 1$.

The most general charge-reversal operator on H_M is $\alpha\mathbb{I}$, with $|\alpha| = 1$.

The most general charge-reversal operator on H_D is

$$(\xi^A, \eta_{A'}) \to \alpha(\bar{\eta}^A, \bar{\xi}_{A'}) \tag{479}$$

with $|\alpha| = 1$.

32. Parity and Time Reversal

The basic idea of the remaining two discrete symmetries — parity and time reversal — is essentially the same as that for charge reversal. One is concerned primarily with finding operators which commute with the S-matrix, and operators which can be interpreted as representing parity and time reversal are particularly good candidates.

We begin with some remarks concerning the Poincaré group. Recall that the restricted Poincaré group, $\mathscr{RP}$, is a connected, 10-dimensional Lie group. This $\mathscr{RP}$ is a normal subgroup of another 10-dimensional Lie group, the full Poincaré group $\mathscr{P}$. However, $\mathscr{P}$ is not connected; it has four connected components. These components consist of Poincaré transformations which reverse neither time nor space orientation ($\mathscr{RP}$), time but not space orientation, space but not time orientation, and both time and space orientation. The quotient group, $\mathscr{P}/\mathscr{RP}$, is isomorphic with the group $\mathbb{Z}_2 \times \mathbb{Z}_2$. ($\mathbb{Z}_2$ is the additive group of integers mod 2.)

The situation is slightly different for the boson and fermion cases. We begin with the boson case. Let H be one of the Hilbert spaces H_{RKG}, H_{CKG}, or H_M. Then, as we have seen (Sect. 16), H defines a representation of $\mathscr{RP}$. That is to say, with each $P \in \mathscr{RP}$ there is associated a unitary operator U_P on H, where these U_P's satisfy:

$$U_P U_{P'} = U_{PP'}$$
$$U_e = \mathbb{I} \tag{480}$$

The problem of obtaining parity-reversal and time-reversal operators can be stated as follows: we wish to extend this representation from $\mathscr{RP}$ to $\mathscr{P}$. That is to say, we wish to find, for each $P \in \mathscr{P}$, a (unitary or antiunitary) operator U_P, subject to (480) and to the condition that, for $P \in \mathscr{RP}$, this representation reduces to the given representation (Sect. 16) of it $\mathscr{RP}$. It is necessary to admit both unitary and antiunitary operators for, as we shall see shortly, it is otherwise impossible to find any extension of our original representation of $\mathscr{RP}$.

There is an important difference between charge reversal on the one hand and parity and time reversal on the other. In the case of charge reversal, one settles eventually on a single unitary charge-reversal operator C. There is, however, no one natural "parity-reversal operator P" or "time-reversal operator T". There is, instead, a 10-dimensional manifold's worth of such operators, namely, the operators associated with the appropriate component of the Poincaré group.

Suppose now that we have a representation of $\mathscr{P}$ as described above. Let P and Q be in the same component of $\mathscr{P}$, so $PQ \in \mathscr{RP}$. Now $U_P U_Q = U_{PQ}$. But, since $PQ \in \mathscr{RP}$, U_{PQ} is unitary rather than antiunitary. We conclude that either both U_P and U_Q are unitary, or else both are antiunitary. (The product of two unitary operators, or two antiunitary operators, is unitary; the product of a unitary and an antiunitary operator is antiunitary.) We conclude that all the operators associated with the Poincaré transformations in a given component of $\mathscr{P}$ are the same (all unitary or all antiunitary.)

In fact, it follows from an argument similar to that used for charge-reversal that U_P is antiunitary if and only if the Poincaré transformation P reverses time. From the remarks above, it suffices to show that, for some P which reverses parity but not time, U_P is unitary, and that, for some P which reverses time but not parity, U_P is antiunitary. Let r^a be a constant, unit, future-directed timelike vector field, and consider the energy operator $\underline{E}$ given by (471). All expectation values of $\underline{E}$ are non-negative. Fix an origin O, and let t^a be a unit, future-directed timelike vector at O. Let P denote the Poincaré transformation which sends

$$x^a \rightarrow x^a + 2t^a(x^b t_b) \tag{481}$$

where x^a is the position vector relative to O. Evidently, this P reverses time orientation but not space orientation. From the commutativity properties of the Poincaré group,

$$U_P^{-1} \underline{E} U_P = \left(U_P^{-1} \frac{i}{\hbar} U_P \right) r'^a \underline{L}_a \tag{482}$$

where

$$r'^a = r^a + 2t^a(r^b t_b) \tag{483}$$

Since r'^a is a past-directed timelike vector, the positivity of $\underline{E}$ implies, evidently, that U_P must be antiunitary. Similarly, let Q be the Poincaré transformation

$$x^a \rightarrow -x^a - 2t^a(x^b t_b) \tag{484}$$

So Q reverses spatial orientation but not time orientation. Then

$$U_Q^{-1} \underline{E} U_Q = \left(U_Q^{-1} \frac{i}{\hbar} U_Q \right) r'^a \underline{L}_a \tag{485}$$

where

$$r'^a = -r^a - 2t^a(r^b t_b) \tag{486}$$

Clearly, in this case r'^a is a future-directed timelike vector, hence U_Q must be unitary.

We next consider uniqueness. Let U_P and U_P' be two extensions of our representation of $\mathscr{RP}$, so $U_P = U_P'$ for $P \in \mathscr{RP}$. Let Q be a Poincaré transformation which reverses, say, temporal orientation but not spatial orientation. Consider the unitary operator

$$A = U_Q' U_Q^{-1} \tag{487}$$

We first how that this A depends only on the component of $\mathscr{P}$ in which Q lies. Let W be another Poincare transformation which lies in the same component of $\mathscr{P}$ as Q. Then $W = QR$ for some $R \in \mathscr{RP}$. Hence,

$$U_W' U_W^{-1} = U_{QR}' U_{QR}^{-1} = (U_Q' U_R')(U_R^{-1} U_Q^{-1}) = U_Q' U_Q^{-1} \tag{488}$$

where we have used the fact that $U_R = U_R'$. We next show that, for $P \in \mathscr{RP}$, U_P commutes with A. Indeed, since $PQ = QV$ for some $V \in \mathscr{RP}$, we have

$$U_P A U_P^{-1} = U_P U_Q' U_Q^{-1} U_P^{-1} = U_{PQ}' U_{PQ}^{-1}$$
$$= U_{QV}' U_{QV}^{-1} = (U_Q' U_V)(U_V^{-1} U_Q^{-1}) = U_Q' U_Q^{-1} = A \tag{489}$$

These properties do not yet suffice, however, to show that A is a multiple of the identity (see p. 132). We must impose an additional condition which ensures that A does not mix up particles and antiparticles. However, it is reasonable, on physical grounds, to make the following additional assumption: U_P reverses the roles of particles and antiparticles if and only if P reverses time directions. Under this assumption, A must be a multiple of the identity, whence $U_Q = \alpha U_Q'$, where α is some complex number (the same for every on Poincaré transformation in the same component as Q). However, since $QQ \in \mathscr{RP}$, we have

$$U_Q' U_Q' = U_{QQ}' = U_{QQ} = U_Q U_Q \tag{490}$$

whence $\alpha = \pm 1$.

To summarize, we are interested in extending a given representation from $\mathscr{RP}$ to $\mathscr{P}$ in such a way that U_P reverses the role of particles and antiparticles when and only when P reverses temporal orientation. Every such extension has the property that U_P is antiunitary when and only when P reverses temporal orientation. The extension of the representation is unique except for the following possibilities: affix a minus sign to U_P whenever P reverses spatial parity, affix a minus sign to U_P whenever reverses temporal orientation, or affix a minus sign to U_P whenever P reverses spatial or temporal orientation, but not both. Thus, from one extension of the representation, it is easy to write down them all.

Finally, we write down a representation of $\mathscr{P}$ for each of our Hilbert spaces H_{RKG}, H_{CKG}, and H_M. Let P be a Poincaré transformation, and write Px for the point of Minkowski space to which P sends the point x. For the real and complex Klein-Gordon cases, the action of U_P on an element $\phi(x)$ of our Hilbert space is as follows:

$$\phi(x) \to \phi(Px) \tag{491}$$

This action clearly defines a representation of $\mathscr{P}$. For the Maxwell case, note that P is a smooth mapping from Minkowski space to Minkowski space, and hence P sends any vector field on Minkowski space to another vector field. This action defines U_P on H_M. For example, the two Poincaré transformations (481) and (484) have the following actions on the vector potential $A_a(x)$:

$$A_a(x) \to A_a(Px) + 2t_a(t^b A_b(Px))$$
$$A_a(x) \to -A_a(Qx) - 2t_a(t^b A_b(Qx)) \tag{492}$$

The situation with regard to H_D differs in some important respects from that above. The fundamental difference is that H_D does not define a representation of $\mathscr{RP}$. Instead, the operative group is what is usually called "inhomogeneous $SL(2,\mathbb{C})$": the (double) covering group of $\mathscr{RP}$. ($SL(2,\mathbb{C})$ is the (double) covering group of the restricted Lorentz group.) We shall denote this connected, 10-dimensional Lie group by $\mathscr{RS}$. If $\mathscr{RS}$ is to replace $\mathscr{RP}$, what group should replace $\mathscr{P}$? One could, in fact, introduce such a group, and attempt to extend to it our representation of $\mathscr{RS}$. It is simpler, however, to proceed in a slightly different way.

Let $P \in \mathscr{RP}$. Associated with P there are precisely two elements of $\mathscr{RS}$. The corresponding pair of unitary operators on H_D differ only in sign. Thus, we can regard H_D as a double-valued representation of $\mathscr{RP}$: with each $P \in \mathscr{RP}$, there is associated two unitary operators on H_D, these operators differing only in sign. These operators satisfy (480), modulo sign. The question in the fermion case is therefore the following: Can this double-valued representation of $\mathscr{RP}$ be extended to a double-valued representation of $\mathscr{P}$? The argument given earlier shows that the operators associated with $P \in \mathscr{P}$ are antiunitary if and only if P reverses time-orientation. The uniqueness situation is essentially the same, by the same argument.

There remains, therefore, only the task of specifying what the operators are for $P \notin \mathscr{RP}$. Since the (double-valued) action of $\mathscr{RP}$ on H_D is known, we need only specify U_P for one time-reversing P and one parity-reversing P. The action is as follows. For the Poincaré transformation (481),

$$(\xi^A(x), \eta_{A'}(x)) \rightarrow \pm \frac{i}{\sqrt{2}} (t^{AA'} \eta_{A'}(Px), t_{AA'} \xi^A(Px)) \qquad (493)$$

and for the Poincaré transformation (484),

$$(\xi^A(x), \eta_{A'}(x)) \rightarrow \pm \frac{1}{\sqrt{2}} (t^{AA'} \eta_{A'}(Qx), t_{AA'} \xi^A(Qx)) \qquad (494)$$

33. Extending Operators to Tensor Products and Direct Sums

We have now introduced a large number of operators — some on the one-particle Hilbert spaces and some of the Fock spacs. In order that such operators can be discussed relative to the S-matrix, however, their action must be defined on the non-interacting Hilbert space $\mathscr{H}$. Since $\mathscr{H}$ arises from two constructions — the tensor product and direct sum of Hilbert spaces — we are led to the problem of extending the action of operators through these two constructions.

We begin with the direct sum. Let $H_1, H_2, \ldots$ be a sequence of Hilbert spaces. Then the direct sum of this sequence, $H = H_1 \oplus H_2 \oplus H_3 \oplus \ldots$ is the Hilbert space consisting of sequences

$$\sigma = (\sigma_1, \sigma_2, \sigma_3, \ldots) \tag{495}$$

with $\sigma_i \in H_i$ for which the sum

$$\|\sigma\|^2 = \|\sigma_1\|^2 + \|\sigma_2\|^2 + \cdots \tag{496}$$

which defines the norm, converges. Now let $O_1, O_2, \ldots$ be a sequence of operators (O_i on H_i) which are either all linear or all antilinear. We then define an operator O on H as follows:

$$O\sigma = (O_1\sigma_1, O_2\sigma_2, O_3\sigma_3, \ldots) \tag{497}$$

Clearly, O is linear (resp., antilinear) provided the O_i are linear (resp., antilinear.) Note, furthermore, that if all the O_i are norm-preserving, so is O; if all the O_i are Hermitian, so is O; if all the O_i are projection operators, so is O. Of course, not every operator on H can be expressed in the form (497).

The tensor product is next. Let $H^\alpha, H^\beta, \ldots, H^\delta$ be a finite sequence of Hilbert spaces. Then the tensor product of this sequence, $H = H^\alpha \otimes H^\beta \otimes \cdots \otimes H^\delta$ is the Hilbert space obtained as the completion of the inner-product space consisting of all formal expressions of the form:

$$\xi^{\alpha\beta\cdots\delta} = \sigma^\alpha \tau^\beta \cdots \mu^\delta + \cdots + \lambda^\alpha \rho^\beta \cdots \nu^\delta \tag{498}$$

137

(The index indicates the Hilbert space to which a vector belongs.) Now let $O^{\alpha'}{}_\alpha, O^{\beta'}{}_\beta, \ldots, O^{\delta'}{}_\delta$ be a sequence of linear operators on $H^\alpha, H^\beta, \ldots, H^\delta$, respectively. We then can define an operator O on H by:

$$O(\xi^{\alpha\beta\cdots\delta}) = (O^{\alpha'}{}_\alpha \sigma^\alpha)(O^{\beta'}{}_\beta \tau^\beta) \cdots (O^{\delta'}{}_\delta \mu^\delta) + \cdots$$
$$+ (O^{\alpha'}{}_\alpha \lambda^\alpha)(O^{\beta'}{}_\beta \rho^\beta) \cdots (O^{\delta'}{}_\delta \nu^\delta) \quad (499)$$

If all the O's are unitary, so is O; if all the O's are Hermitian, so is O; if all the O's are projection operators, so is O. Now suppose that the given O-operators are antilinear rather than linear. We thus have linear mappings from $\bar{H}_\alpha$ to H^α, from $\bar{H}_\beta$ to H^β, etc. Thus, in the index notation, the O's would be written $O^{\alpha'\alpha}, O^{\beta'\beta}, \ldots, O^{\delta'\delta}$. In this case, we define the corresponding operator O on H by

$$O(\xi^{\alpha\beta\cdots\delta}) = (O^{\alpha'\alpha}\bar{\sigma}_\alpha)(O^{\beta'\beta}\bar{\tau}_\beta) \cdots (O^{\delta'\delta}\bar{\mu}_\delta) + \cdots$$
$$+ (O^{\alpha'\alpha}\bar{\lambda}_\alpha)(O^{\beta'\beta}\bar{\rho}_\beta) \cdots (O^{\delta'\delta}\bar{\nu}_\delta) \quad (500)$$

If the O's are anti-unitary, so is O.

We next consider the application of these constructions to obtaining operators on $\mathscr{H}$. Recall that the non-interacting Hilbert space $\mathscr{H}$ is the tensor product of certain Fock spaces based on one-particle Hilbert spaces, e.g.,

$$\mathscr{H} = \mathscr{F}(H_D) \otimes \mathscr{F}(H_M) \otimes \mathscr{F}(H_{CKG}) \quad (501)$$

where $\mathscr{F}$ denotes the operation of taking the (symmetric or antisymmetric, as appropriate) Fock space.

Consider first the unitary or antiunitary operators U_P ($P \in \mathscr{P}$) which arise from the Poincaré group. These operators are defined originally on the one-particle Hilbert spaces. Their action is first extended to the many-particle Hilbert spaces via (499) or (500), and then to the Fock spaces via (497). Finally, these operators are defined on $\mathscr{H}$ via (499) or (500). Thus, we obtain a representation of the Poincaré group $\mathscr{P}$ on $\mathscr{H}$. For $P \in \mathscr{P}$, we write the corresponding operator on $\mathscr{H}$ as U_P. (No confusion will result from this duplicity of notation.) Note that all the operators U_P on $\mathscr{H}$ are norm-preserving, and that U_P is antilinear if and only if P reverses time orientation, linear otherwise. The energy, momentum, and angular momentum operators on $\mathscr{H}$ are obtained by considering the U_P's which differ infinitesimally from the identity (see Eqn. (222).) Similarly, the charge-reversal operator C is defined, first, on the one-particle Hilbert spaces, and then extended successively to the many-particle spaces, to the Fock spaces, and finally to $\mathscr{H}$. The resulting operator on $\mathscr{H}$ is again denoted by C.

Another operator of interest is the total charge operator, Q. On our real Hilbert spaces (which represent neutral particles), H_{RKG} and H_M, $Q = 0$. On the complex Hilbert spaces, H_{CKG} and H_D, Q takes one of the two forms

$$Q = eP^- - eP^+ \quad (502)$$

$$Q = eP^+ - eP^- \tag{503}$$

where P^+ and P^- are the projection operators onto positive-frequency and negative-frequency parts, respectively. The choice between (502) or (503) is based essentially on convention. One uses (502) when the "particles" (positive-frequency solutions) have negative charge, and (503) when they have positive charge. (For example, one could perfectly well dictate that the positron is the "particle", and the electron the antiparticle. It is conventional, however, to make the assignments the other way. This choice his no physical consequences.) Note that, in every case, Q is Hermitian. We extend Q from the one-particle to the many-particle Hilbert spaces via (499), to the Fock spaces via (497), and to $\mathscr{H}$ via (499). The result is a Hermitian charge operator, Q, on $\mathscr{H}$.

Finally, we consider the creation, annihilation, and field operators. These operators, in distinction to the others defined above, are first defined on the Fock spaces rather on the one-particle Hilbert spaces. Suppose, for example, that $\mathscr{H}$ is given by (501), and we are considering the field operator $\underline{\phi}(f)$ on $\mathscr{F}(H_{CKG})$. Now consider the following triple of operators: $\mathbb{I}$ (the identity) on $\mathscr{F}(H_D)$, $\mathbb{I}$ on $\mathscr{F}(H_M)$, and $\underline{\phi}(f)$ on $\mathscr{F}(H_{CKG})$. This triple defines, via the construction (499), an operator on $\mathscr{H}$. In this way, the creation, annihilation, and field operators are extended from a single Fock space to $\mathscr{H}$. The resulting operators on $\mathscr{H}$ will, as usual, be denoted by the same symbols as the corresponding operators on the single Fock spaces.

All of the commutators and other relations between these various operators on $\mathscr{H}$ are simple to evaluate. We give a few examples. The operators associated with the Poincaré group satisfy, of course,

$$U_P U_{P'} = U_{PP'} \tag{504}$$

The U_P's leave invariant the charge of an element of $\mathscr{H}$ if P does not reverse the orientation, and reverse that sign if P does reverse time orientation. Hence, we have

$$U_P Q U_P^{-1} = \pm Q \tag{505}$$

with the plus sign if P does not reverse time orientation, the minus sign otherwise. Of course, charge-reversal reverses the sign of charge:

$$CQC^{-1} = -Q \tag{506}$$

The commutators between the field operators (and the creation and annihilation operators) within one Fock space are the same for $\mathscr{H}$ as for the original Fock space. That is, for example, we have

$$\left[\underline{\phi}(f), \underline{\phi}^*(g)\right] = \frac{\hbar}{2i} D(f,g)\mathbb{I} \tag{507}$$

on $\mathscr{H}$. Similarly, the adjoint relations between these operators are unchanged in the passage to $\mathscr{H}$. Field operators (as well as creation and annihilation operators) which act on different Fock spaces commute. For example,

$$\left[\underline{\phi}(f), \underline{A}(f^a)\right] = 0 \tag{508}$$

Finally, we consider the relationship between the field operators and C, Q, and U_P. Once again, everything is straightforward, so a single example will suffice. Consider H_{CKG}, so

$$\underline{\phi}(f) = \hbar C(\sigma^-(f)) + \hbar A(\overline{\sigma^+(f)})$$
$$\underline{\phi}^*(f) = \hbar A(\overline{\sigma^-(f)}) + \hbar C(\sigma^+(f)) \tag{509}$$

Let P be a restricted Poincaré transformation. Then, from (497), (499), (100), and (101),

$$U_P\underline{\phi}(f)U_P^{-1} = \hbar C(U_P\sigma^-(f) + \hbar A(\overline{U_P\sigma^+(f)}) \tag{510}$$

Similarly, if C is a particle-reversal operator, then

$$C\underline{\phi}(f)C^{-1} = \hbar C(C\sigma^-(f)) + \hbar A(\overline{C\sigma^+(f)})$$
$$= \hbar C(\sigma^+(f)) + \hbar A(\sigma^-(f)) = \underline{\phi}^*(f) \tag{511}$$

For the charge operator Q, we note that $\underline{\phi}(f)$ creates an antiparticle (say, with positive charge) and annihilates a particle. Hence, the total change in the charge effected by $\underline{\phi}(f)$ is just $-e$ times the norm of $\sigma^-(f)$ plus $-e$ times the norm of $\sigma^+(f)$. Thus,

$$[Q, \underline{\phi}(f)] = -e\left(\|\sigma^+(f)\| + \|\sigma^-(f)\|\right)\mathbb{I} \tag{512}$$

Clearly, the list of operators in this subject is almost infinite. Roughly speaking, any two operators in this list have a relationship which is simple, straightforward to derive, and easy to interpret physically.

34. Electromagnetic Interactions

In Sect. 29 (see Eqn. (455)) we wrote down an expression for the S-matrix in terms of an (unknown) operator field $\underline{K}(x)$ on Minkowski space. Of course, this formula gives practically no information about the scattering unless one knows $\underline{K}(x)$. One imagines that the actual $\underline{K}(x)$ which describes physical processes in the real world can be written as the sum a certain number of terms (e.g., the electron-photon interaction, the nucleon-photon interaction, the π-meson-photon interaction (electromagnetic interactions), the π-meson-nucleon interaction (strong interactions), the electron-neutrino interaction, the π-meson-neutrino interaction (weak interactions), etc.) There are at least some experimental situations in which one single term dominates all the others. One attempts to obtain an expression for this term using physical arguments and trial and error. That is to say, one makes a reasonable guess for the term in $\underline{K}(x)$, and compares the theoretical consequences of that guess (via (455)) with experiment. The hope is that one can, in this way, isolate and study each term, and then, by adding the well-established terms, obtain a reasonable approximation to the "actual" $\underline{K}(x)$ which is operative in Nature. We shall here merely illustrate the general idea by writing down and discussing a few of the $\underline{K}(x)$'s associated with the interaction of charged particles with the electromagnetic field.

We begin with the simplest case: the interaction of a complex Klein-Gordon field with the Maxwell field, e.g., the interaction of $\pi^{\pm}$-mesons with photons. In this case, we would take for our non-interacting Hilbert space

$$\mathscr{H} = \mathscr{F}(H_{CKG}) \otimes \mathscr{F}(H_M) \tag{513}$$

What should we choose for the operator field $\underline{K}(x)$ on $\mathscr{H}$? In this case, we have an important physical clue: we know what the classical "interaction energy density" is between a classical Klein-Gordon field and a classical Maxwell field, namely

$$K(x) = \frac{ie}{2\hbar} \left(\phi(x) \nabla_a \bar{\phi}(x) - \bar{\phi}(x) \nabla_a \phi(x) \right) A^a(x) \tag{514}$$

(Note that it is only an integral of (514) which has meaning, for we have the freedom to add a gradient to the vector potential. Appropriate integrals are gauge-invariant, however, because the first integral is, as we have seen, divergence-free.) In (514), e is a constant. Using the discussion of Sect. 30, and the fact that $\underline{K}(x)$

141

has dimensions of energy density (mt^{-3}), we see that the coupling constant e has dimensions $m^{1/2}t^{1/2}$, whence $e^2/\hbar$ is dimensionless. In order to obtain eventual agreement with experiment, it will, of course, be necessary to set this constant to $1/137$.

While (514) is a perfectly nice scalar field (on Minkowski space) constructed out of a complex Klein-Gordon field and a Maxwell field, it is, unfortunately, just that — a scalar field rather than an operator field. The expression (514) is just not the right sort of mathematical object to be $\underline{K}(x)$. Now comes the "transition from classical to quantum theory." Roughly speaking, what we propose to do is to replace the classical fields in $K(x)$ (Eqn. (514)) by the corresponding field operators to obtain $\underline{K}(x)$. Unfortunately, this replacement is not so simple and unambiguous as it may appear at first sight.

By what operator should we replace $\phi(x)$? Our Klein-Gordon field operator, $\underline{\phi}(f)$, depends on test fields in Minkowski space, and not on points of Minkowski space. What one would like to do is define an operator field $\underline{\phi}(x)$ by

$$\underline{\phi}(x) = \lim_{f \to \delta_x} \underline{phi}(f) \tag{515}$$

where δ_x denotes a δ-function located at the point x. But will the limit in (515) exist? The answer, as we have seen earlier, is no. We could still regard $\underline{\phi}(x)$ as an operator-valued distribution (i.e., a linear mapping from test functions to operators on $\mathcal{H}$ — that, after all, is what $\underline{\phi}(f)$ is), but such an attitude again leads to difficulties. Eqn. (514) will require us to take products of such operator-valued distributions, but the ability to take products is precisely what is lost in the transition from functions to distributions. That is to say, products of distributions are not in general well-defined. (This is a genuine and serious problem — not to be confused, for example with the standard complaints about use of the Dirac δ-function.) In short, we are stuck. There is no meaning which can be given to (515) which would be appropriate for replacement in (514).

We adopt the following attitude. We leave the problem of the nonexistence of limits such as (515) unresolved for the time being. We permit ourselves to manipulate such quantities formally, as though the question of the limits had never arisen. This marks the third (and, mercifully, the last) of the mathematical problems associated with this formalism. For emphasis, we list these problems again:

1. The question of the convergence of the infinite sun of operators in (455).

2. The question of the convergence of the integrals (over all of Minkowski space) of operators in (455).

3. The nonexistence of the "δ-function limits" of field operators used in obtaining $\underline{K}(x)$.

The situation will look less gloomy shortly. (I find it hard to believe that the "ultimate", mathematically acceptable, quantum field theory will result from a brute-force attack on these problems.)

We now have $\underline{\phi}(x)$ to replace $\phi(x)$. Naturally, we replace the classical complex-conjugate field, $\bar{\phi}(x)$, by the adjoint operator, $\underline{\phi}^*(x)$. What about the derivative terms? Let p^a be a vector at the point x, and extend p^a to a constant vector field on Minkowski space. We then define

$$p^a \nabla_a \underline{\phi}(x) = - \lim_{f \to \delta_x} \underline{\phi}(p^a \nabla_a f) \tag{516}$$

(See Eqn. (145).) Similar remarks concerning non-existence of limits apply.

We must next select an operator to replace the vector potential, $A_a(x)$, in (514). Ideally, one would like to define $ulA^a(x)$ by

$$p^a \underline{A}_a(x) = \lim_{f \to \delta_x} \underline{A}(p^a f) \tag{517}$$

where p^a is a constant vector field, and $\underline{A}(\ )$ is the field operator (Sect. 14) for the vector potential. Unfortunately, this won't work, for $\underline{A}(f^a)$ is only defined for test fields f^a which can be written as the sum of a divergence-free field and a gradient; $p^a f$ cannot be written in this form in general. The simplest way of overcoming this difficulty is as follows. First note that the commutator of the vector potential operators ((385) and (386)) is well-defined whether or not the test fields, f^a and g^a, can be written as the sum of a divergence-free field and a gradient. In fact, it is only these commutators which will enter the S-matrix. Hence, we can work with vector potential operators, $\underline{A}_a(x)$, and use for the commutators (385).

We now have an operator equivalent for each term in (514). We must now face the next problem: in what order should the operators be placed? This difficulty does not arise, of course, in the classical theory, because the classical fields may be placed in any order. We consider the most general linear combination:

$$\underline{K}(x) = \frac{ie}{2\hbar} \left(a\underline{\phi}\nabla_a\underline{\phi}^* + b(\nabla_a\underline{\phi}^*)\underline{\phi} + c\underline{\phi}^*\nabla_a\underline{\phi} + d(\nabla_a\underline{\phi})\underline{\phi}^* \right) \underline{A}^a \tag{518}$$

where a, b, c, and d are real numbers. Taking the Hermitian conjugate of (518),

$$\underline{K}(x) = -\frac{ie}{2\hbar} \left(a(\nabla_a\underline{\phi})\underline{\phi}^* + b\underline{\phi}^*\nabla_a\underline{\phi} + c(\nabla_a\underline{\phi}^*)\underline{\phi} + d\underline{\phi}(\nabla_a\underline{\phi}^*) \right) \underline{A}^a \tag{519}$$

We see that the Hermiticity of (518) requires

$$a = -d \qquad b = -c \tag{520}$$

Further information about the coefficients is obtained from the experimental fact that electromagnetic interactions are invariant under charge reversal. From (518):

$$C\underline{K}(x)C^{-1} = \frac{ie}{2\hbar} \left(a\underline{\phi}^*\nabla_a\underline{\phi} + b(\nabla_a\underline{\phi})\underline{\phi}^* + c\underline{\phi}\nabla_a\underline{\phi}^* + d(\nabla_a\underline{\phi}^*)\underline{\phi} \right) \underline{A}^a \tag{521}$$

Thus, invariance under charge reversal requires one of the following two alternatives:

$$a = c \qquad\qquad b = d \qquad\qquad C\underline{A}^aC^{-1} = \underline{A}^a \tag{522}$$

$$a = -c \qquad\qquad b = -d \qquad\qquad C\underline{A}^aC^{-1} = -\underline{A}^a \tag{523}$$

We choose (523) for two reasons: (i) it is more reasonable on physical grounds to have the vector potential reverse sign under charge reversal (for classical electromagnetic fields reverse sign when the signs of all charges are reversed), and (ii) with this choice, $\underline{K}(x)$ reduces, in the classical limit, to the classical expression (514). Thus, we arrive at the interaction:

$$\underline{K}(x) = \frac{ie}{4\hbar} \left(\underline{\phi} \nabla_a \underline{\phi}^* + (\nabla_a \underline{\phi}^*) \underline{\phi} - \underline{\phi}^* \nabla_a \underline{\phi} - (\nabla_a \underline{\phi}) \underline{\phi}^* \right) \underline{A}^a \qquad (524)$$

Eqn. (524) describes an interaction which is invariant under charge reversal. Note, furthermore, that if P is any Poincaré transformation, then

$$U_P \underline{K}(x) U_P^{-1} = \underline{K}(Px) \qquad (525)$$

Since $\underline{K}(x)$ is integrated over all of Minkowski space (Eqn. (455)), the final S-matrix will commute with each U_P. Thus, our interaction conserves the quantities associated with the infinitesimal generators of energy, momentum, and angular momentum. The interaction is also invariant under parity and time reversal. Note, furthermore, that we have

$$[\underline{K}(x), \underline{K}(y)] = 0 \qquad (526)$$

for $x - y$ spacelike, for when $x - y$ is spacelike, any two operators in (524) commute with each other. Finally, the total charge operator, Q, commutes with $\underline{K}(x)$, for $\underline{A}_a(x)$ commutes with Q, $\underline{\phi}(x)$ decreases the total charge by $2e$, while $\underline{\phi}^*(x)$ increases it by $2e$.

To summarize, (524) is an interaction which is invariant under parity, time, and charge reversal, and conserves charge, energy, momentum, and angular momentum.

As a second example, we discuss the interaction of photons with electrons and positrons. As before, we begin with the classic interaction energy density:

$$K(x) = e \left(\xi^A \bar{\xi}^{A'} + \eta^{A'} \bar{\eta}^A \right) A_{AA'} \qquad (527)$$

The classical Dirac fields are to be replaced by the following operators:

$$p_A \underline{\xi}^A(x) = \lim_{f \to \delta_x} \frac{1}{2} \left[\underline{\psi}(fp^A, f\bar{p}_{A'}) - i\underline{\psi}(ifp^A, -if\bar{p}_{A'}) \right]$$
$$\bar{p}^{A'} \underline{\eta}_{A'}(x) = \lim_{f \to \delta_x} \frac{1}{2} \left[\underline{\psi}(fp^A, f\bar{p}_{A'}) + i\underline{\psi}(ifp^A, -if\bar{p}_{A'}) \right] \qquad (528)$$

where p^A is a constant spinor field. Note that e in (527) again has dimensions $m^{1/2}t^{1/2}$. The classical complex-conjugate fields, $\bar{\xi}^{A'}$ and $\bar{\eta}_A$, are to be replaced by the Hermitian conjugates, $\underline{\xi}^{*A'}$ and $\underline{\eta}_A^*$, respectively. In this case, the problem of factor ordering is not resolved by the requirement that $\underline{K}(x)$ be Hermitian: this condition is satisfied for any factor ordering. However, this electromagnetic interaction should be invariant under charge reversal. We have

$$C\underline{\xi}^A C^{-1} = \underline{\eta}^{*A}$$
$$C\underline{\eta}_{A'} C^{-1} = \underline{\xi}^*_{A'} \qquad (529)$$

What should we adopt for the behavior of the vector potential operator, $\underline{A}_a(x)$, under charge reversal? We have already decided, for the meson-photon interaction, to use

$$C\underline{A}_a C^{-1} = -\underline{A}_a \qquad (530)$$

It is an important point that we must choose the same behavior for the present interaction. The reason is that, for the "actual" interaction which Nature obeys, $\underline{K}(x)$ will be the sum of the various interactions. If we use a different charge-reversal operator for each term which appears in this sum, then we will have no operator which commutes with the total $\underline{K}(x)$. In other words, the behavior of each type of particle under the various reversals must be fixed once and for all. One has, of course, freedom to choose that behavior, and this choice is based on obtaining operators which commute with as many terms in the "final" $\underline{K}(x)$ possible. Thus, using (529) and (530), we are led to adopt the expression

$$\underline{K}(x) = \frac{1}{2}e\left(\underline{\xi}^A \underline{\xi}^{*A'} - \underline{\xi}^{*A'}\underline{\xi}^A + \underline{\eta}^{A'}\underline{\eta}^{*A} - \underline{\eta}^{*A}\underline{\eta}^{A'}\right)\underline{A}_{AA'} \qquad (531)$$

for the interaction.

Note that (530) is Hermitian, and that the resulting S-matrix commutes with the unitary operators which arise from the Poincaré group. Thus, (530) conserves energy, momentum, and angular momentum. By the same argument as before, $\underline{K}(x)$ commutes with the total charge operator Q. Finally, we note that, if $x - y$ is spacelike,

$$[\underline{K}(x), \underline{K}(y)] = 0 \qquad (532)$$

This arises from the following facts: when $x - y$ is spacelike, any two boson operators commute, while any two fermion operators anticommute. But $\underline{K}(x)$ contains an even number of fermion operators. Since reversing the order of two boson operators gives a plus sign, and reversing the order of two fermion operators gives a minus sign, the total number of minus signs will be even, and so we have (532).

Clearly a vast number of conceivable interactions could be written down using the pattern illustrated above. One first writes down a real scalar field constructed from the classical fields. One then replaces the classical fields by the corresponding operators. The factors must be ordered so that the resulting operator is Hermitian, and satisfies (532). Beyond that, the choice of factor ordering must be based on physical or aesthetic considerations, experiment, etc. We have merely discussed two possible interactions here in order to illustrate the method. (In fact, these are the two simplest, for one can rely heavily on classical theory as a guide.)

35. Transition Amplitudes

Suppose now that we have selected a particular $\underline{K}(x)$, and wish to work out its experimental consequences, using (455). The straightforward procedure — substituting $\underline{K}(x)$ into (455), and attempting to carry out the integrals and sum — turns out to be too difficult to carry out in practice. Instead, one adopts a more indirect approach — which leads ultimately to the Feynman rules. We shall not attempt to derive the Feynman rules, or even discuss the large volume of technical apparatus which has been developed to deal with (455). Instead, we merely indicate the general idea of the method.

Suppose first that we were able, in some way, to obtain the value of the complex number

$$(\tau, S\sigma) \tag{533}$$

for any two states $\sigma, \tau \in \mathscr{H}$. (The expression (533) is called the *transition amplitude* between the state σ and the state τ.) This information is, of course, completely equivalent to a knowledge of the S-matrix. In fact, it suffices to know (533) only for σ's and τ's drawn from a certain subspace of $\mathscr{H}$, provided this subspace is dense in $\mathscr{H}$. Let σ_0 denote the vacuum state in $\mathscr{H}$ and $C_1, C_2, \ldots, C_n$ any finite sequence of creation operators on $\mathscr{H}$. (One C_i might create a photon, another an electron, others mesons, etc.) We consider the state (element of $\mathscr{H}$)

$$C_1 C_2 \cdots C_n \sigma_0 \tag{534}$$

Clearly, the collection of all finite linear combinations of states of the form (534) is dense in $\mathscr{H}$. Hence, it suffices to evaluate

$$(C_1 \cdots C_n \sigma_0, S C_1' \cdots C_m' \sigma_0) = (C_1 \cdots C_n \sigma_0, C_1' \cdots C_m' \sigma_0)$$

$$+ \left(-\frac{i}{\hbar}\right) \int dV_1 (C_1 \cdots C_n \sigma_0, \underline{K}(x_1) C_1' \cdots C_m' \sigma_0)$$

$$+ \frac{1}{2!} \left(-\frac{i}{\hbar}\right)^2 \int dV_1 \int dV_2 (C_1 \cdots C_n \sigma_0, T\left[\underline{K}(x_1), \underline{K}(x_2)\right] C_1' \cdots C_m' \sigma_0)$$

$$+ \cdots \tag{535}$$

for any $C_1, \ldots, C_n, C_1', \ldots, C_m'$. One now attempts to evaluate the various terms in the sum (535) individually. The first term is celled the 0th-order interaction. It clearly vanishes unless $n = m$, and the C's and C''s create the same number of photons, the same number of electrons, etc. The second term is called the first-order interaction, etc. These orders correspond to the various orders of the Feynman diagrams.

The next step is based on the following observation. The operator $\underline{K}(x)$ is expressed in terms of the field operators, and can, therefore, be written in terms of creation and annihilation operators. Thus, each inner product in (535) is equal to the vacuum expectation value of some product of creation and annihilation operators. But we have seen (p. 112) that such expectation values can be reduced to products of the propagators. (One pushes the annihilation operators to the the right, using the commutators, until they reach σ_0, and there give zero.) Thus, each term in (535) can be written as an integral of a product of operators. The Feynman rules represent a technique for writing out such integrals directly without going through the algebra (p. 112). Since $\underline{K}(x)$ was expressed in terms of the "unsmeared" field operators, the propagators will be distributions. Thus, the evaluation of the S-matrix reduces to integrating certain products of distributions over Minkowski space. Of course, the integrals diverge. These are the divergences.

We illustrate these remarks with one example. Consider the second-order interaction with $\underline{K}(x)$ given by (531). Now, $\underline{K}(x)$ contains two lepton operators and one photon operator. Thus, the transition probability will vanish (in the second order) unless the number of outgoing leptons differs by no more than four from the number of ingoing leptons, and the number of outgoing photons differs by no more than two from the number of ingoing photons. Clearly, the higher the order of the interaction, the more possibilities for creating and annihilating particles. Because of the smallness of the coupling constant in this electromagnetic interaction, $e^2/\hbar = 1/137$, many experiments are adequately described by the first few orders.

> "Would it not be better to get *something* done, even though one might not quite understand *what?*"
>
> – J. L. Synge

About the author

Robert Geroch is a theoretical physicist and professor at the University of Chicago. He obtained his Ph.D. degree from Princeton University in 1967 under the supervision of John Archibald Wheeler. His main research interests lie in mathematical physics and general relativity.

Geroch's approach to teaching theoretical physics masterfully intertwines the explanations of physical phenomena and the mathematical structures used for their description in such a way that both reinforce each other to facilitate the understanding of even the most abstract and subtle issues. He has been also investing great effort in teaching physics and mathematical physics to non-science students.

Robert Geroch with his dog Rusty

Made in the USA
Lexington, KY
15 October 2015